Euclid's Elements Bk I and II

by

David B Hayes

Based on the original English Edition prepared by H. Billingsley of London, England, in 1570. The text was compared with the original Greek text and Latin translation prepared by I. L. Heiberg, 1883. The text was compared with the English translation of Sir Thomas Heath (1908 and 1925) and with other translations. Edited and adapted by Dr. David B Hayes. (See Appendix for descriptions of adaptations.)

Principle Source Publisher
Fidelity to Primary Sources
Service to Creative Thinkers

Published by Principal Source Publisher
1998, 1999, 2022

Table of Contents

Preface

This is the second edition of a text of the first two books of Euclid's Elements prepared by David Hayes. The first edition was not formally published but used at Magdalen College in Warner, NH. It was copyrighted in 1998 with typos and corrections copyrighted in 1999. The first edition did not include diagrams for the propositions. As an experiment, the students were instructed to draw the diagrams from the words by themselves. This second edition includes diagrams for every proposition. The text of the propositions is based on the original English Edition prepared by H. Billingsley of London, England, in 1570. The text was compared with the original Greek text and Latin translation prepared by I. L. Heiberg, 1883. The text was compared with the English translation of Sir Thomas Heath (1908 and 1925) and with other translations. Edited and adapted by Dr. David B Hayes. See the Appendix for descriptions of adaptations.

The original message of the translator H. Billingsley of London serves as the main introduction to this book.

The Translator to the Reader. (Billingsley)

There is (gentle reader) nothing (the Word of God only set apart) which so much beautifies and adorns the soul and mind of man, as the knowledge of good arts and sciences; as the knowledge of natural and moral philosophy. The one sets before our eyes the creatures of God, both in the heavens above, and in the earth beneath: in which as in a mirror, we behold the exceeding majesty and wisdom of God, in adorning and beautifying them as we see: in giving unto them such wonderful and manifold properties, and natural workings, and that so diversely and in such variety: further in maintaining and conserving them continually, whereby to praise and adore him, as by St. Paul we are taught. The other teaches us rules and precepts of virtue, how, in common life amongst men we ought to walk uprightly: what duties pertain to our fellows, what pertains to the government or good order both of a household, and also of a city or commonwealth. The reading likewise of histories, conduces not a little to the adorning of the soul and mind of man, a study by all men commended: by it are known the arts and doings of indefinite wise men gone before us. In histories are contained indefinite examples of heroic virtues to be by us followed, and horrible examples of vices to be by us eschewed. Many other arts also there are which beautify the mind of man: but of all other none does more garnish and beautify it than those arts which are called Mathematical. Unto the knowledge of which no man can attain without the perfect knowledge and instruction of the principles, grounds, and Elements of Geometry. But perfectly to be introduced into them requires diligent study and reading of old ancient authors. Among which, none for a beginner is to be preferred before the most ancient philosopher Euclide of Megara. For of all other he hath in a true method and just order gathered together whatsoever any before him had of the elements written: inventing also and adding many things of his own: whereby he hath in due form accomplished (perfected) the art: first giving definitions, principles, and grounds, from which he deduces his propositions or conclusions, in such wonderful ways, that that which goes before is of necessity required to the proof of that which follows. So that without the diligent study of Euclid's Elements, it is impossible to attain to the perfect knowledge of Geometry, and consequently of any of the other mathematical sciences. Wherefore, considering the want and lack of such good authors before now in our English tongue, lamenting also the negligence, and lack of zeal to their country in the life of our nation, to whom God has given both knowledge and also ability to translate into our tongue, and to publish abroad such good authors and books (the chief instruments of all learning): seeing moreover that many good writers both of gentlemen and of others of all degrees, much desirous and studious of the arts, and seeking for them as much as they can, sparing no pains, and yet frustrated in their intent, by no means attaining to that which they seek: I have for their sakes, with some charge and great travail, faithfully translated into our vulgar tongue, and set abroad in print, this book of Euclid. Whereunto I have added easy and plain declarations and examples by figures of the definitions. In which book also you shall find manifold additions, scholium, annotations, and inventions: which I have gathered out of the most famous and chief mathematicians, both of old time and in our age: as by diligent reading it in course you shall well perceive. The fruit and gain which I require for these my pains and travail shall be nothing else, but only that thou gentle reader, will gratefully accept the same: and that thou may thereby receive some profit: and moreover to excite and stir up others learned to do the like, and to take pains in that behalf. By means whereof, our English tongue shall no less be enriched with good authors, than are other strange tongues: as the Dutch, French, Italian, and Spanish: in which are read all good authors in this manner, found amongst the Greeks or Latins. Which is the chief cause, that amongst them does flourish so many cunning and skillful men, in the invention of strange and wonderful things, as in these our days we see them do. Which fruit and gain if I attain unto, it shall encourage me hereafter, in such like sort to translate, and set abroad some other good authors, both pertaining to religion (as partly I have already done) and also pertaining to the Mathematical Arts. Thus gentle reader farewell.

This edition of Euclid's Elments was prepared to be available for study by students who desire to read the Elements with the text being intended to be as close as possible to what Euclid wrote and to what people such as H. Billingsley, Blaise Pascal, George Stanciu, and Bertand Russell read.

George Stanciu, in an essay entitled "Patrick J. Deneen's Res Idiotica and Liberal Education," explains that mathematical truth is appropriate for beginning education because its beauty can strike, pierce, and transform us:

Beauty Can Strike, Pierce, and Transform Us

Kalkavage points out that what is objectively beautiful is intelligible, and, of course, he gives music and Pythagorean tuning as an example, but he could have given all theoretical physics. For me and for Kalkavage, beauty is an essential element of education: "Beauty can take us by surprise. It strikes, pierces, even transforms us."[9]

In the tenth grade, I was transformed by Euclid's demonstration that the prime numbers are infinite, an elegant proof that surprisingly showed in six lines of text an eternal truth. Until that point in my life, I thought truth did not exist; everything about me changed, the seasons, my body, and people. My experience of the human world was that everything was in flux, sometimes bordering on the absurd. Suddenly, mathematics presented me with one thing that was unchanging, a timeless truth; the 2,500 years between Euclid and me disappeared. My encounter with Euclid was not unique. Bertrand Russell described his first encounter with serious mathematics: "At the age of eleven, I began Euclid, with my brother as my tutor. This was one of the great events of my life, as dazzling as first love. I had not imagined that there was anything so delicious in the world."[16]

(Quote from: George Stanciu, "Patrick J. Deneen's Res Idiotica and Liberal Education," (2021), https://allthelightwecansee.com/patrick-j-deneens-residiotica-and-liberal-education/. The references in Stanciu's text are: [9] Peter Kalkavage, all quotations are from The Neglected Muse: Why Music Is An Essential Liberal Art, The Imaginative Conservative (Feb. 21, 2016); and [16] Bertrand Russell, The Autobiography of Bertrand Russell: 1872-1914 (Boston: Atlantic Monthly Press, 1967), pp. 37-38.)

Modern philosophers and mathematicians may argue about whether or not mathematical proofs are discovered or invented. However, the Greeks, including Euclid, unequivocally thought of mathematics as a part of eternal truth that is discovered. The use of sentences such as "let it have been postulated, from every point to every point, to draw a straight line," leaves no doubt that Euclid is not claiming to have invented this statement. Plato's Academy is reported to have a sign posted over the door that said, "Let no one ignorant of geometry enter." At the Academy, universal questions of interest to every human being, such as "Who am I?" were asked. But Plato believed that the benefits of studying geometry were a necessary preparation for further study in philosophy. That trustworthy truthfulness is contained in demonstrations such as Euclid's Elements is also attested by the ancient philosopher Aristotle:

Aristotle, Posterior Analytics 11. 100b5-14:

Some of the intellectual characteristics by which we attain truth are always truthful, while others, such as opinion and calculation, admit of falsehood. Now, since (a) scientific knowledge and intelligence are always truthful, and no other kind (of intellectual characteristic) is more precise than scientific knowledge -- except for intelligence; (b) fundamental principles are more knowable than demonstrations, and (c) all scientific knowledge involves reasoning; (it follows that) scientific knowledge does not apprehend fundamental principles, and since only intelligence can be more truthful than scientific knowledge, it is intelligence which apprehends the fundamental principles. This result also follows from the fact that there cannot be demonstration of the fundamental principle of demonstration, nor, consequently, scientific knowledge of scientific knowledge.

(Translation from the footnote in The Nicomachean Ethics, Martin Ostwald translation, Prentice Hall, 1990)

The copious notes, skeptical discussions, additions and corrections to Euclid with the careful scholarly tracing of who invented what proof first in human history is judged, by this author, to be trivial compared to the trustworthy truth presented by Euclid with the propositions chosen and ordered properly. Beginning students are advised to begin reading Euclid with this simple edition before proceeding to the more complicated annotated editions.

The Elements Book I

Definitions:

1. A point is that which has no part.
2. A line is length without breadth.
3. The extremities of lines are points.
4. A straight line is that which lies equally between its points.
5. A surface is that which has only length and breadth.
6. The extremities of a surface are lines.
7. A plane surface is that which lies equally between its straight lines.
8. A plane angle is an inclination of two lines in a plane, the one to the other, and the one touching the other and not being joined in a straight line.
9. And if the lines which contain the angle be straight, then it is called a rectilineal angle.
10. When a straight line standing upon a straight line makes the adjacent angles equal, then either of those angles is a right angle. And the straight line that stands erect is called a perpendicular to that upon which it stands.
11. An obtuse angle is that which is greater than a right angle.
12. An acute angle is that which is less than a right angle.
13. A boundary is the extremity of any thing.
14. A figure is that which is contained by one boundary or many boundaries.
15. A circle is a plane figure contained by one line, which is called the circumference, unto which all straight lines drawn from one point within the figure and falling upon the circumference thereof are equal to one another.
16. And that point is called the center of the circle.
17. A diameter of a circle is a straight line, which drawn through the center thereof and ending on the circumference on either side, divides the circle into two equal parts.
18. A semicircle is a figure contained by the diameter and by that part of the circumference that is cut off by the diameter.
19. Rectilineal figures are those that are contained by straight lines: trilateral by three, quadrilateral by four, multilateral by more than four straight lines.
20. Of trilateral figures, an equilateral triangle is that which has three equal sides, an isosceles triangle has only two sides equal, a scalene triangle has all three sides unequal.
21. Again, of triangles, a right angled triangle has a right angle, an obtuse angled triangle has an obtuse angle, an acute angled triangle has all three of its angles acute.
22. Of quadrilateral figures, a square is that whose sides are equal and its angles are right

angles; an oblong is that which has right angles, but not equal sides; a rhombus has four equal sides, but is not right angled; a rhomboid has opposite sides equal and its opposite angles are equal, but it has neither equal sides nor right angles; all other figures with four sides besides these are called trapezia.

23 Parallel lines are straight lines that being in one and the same plane, and extended indefinitely on both sides, do not meet.

Postulates:

1. Let it have been postulated: from every point to every point, to draw a straight line;
2. and to extend every straight line in a straight line indefinitely;
3. and upon every center and every distance to describe a circle;
4. and all right angles are equal to one another;
5. and when a straight line falling upon two straight lines makes on one and the same side, the two interior angles less than two right angles, then shall these two straight lines being extended indefinitely meet on that side in which are the two angles less than two right angles.

Common Notions:

1. Things equal to the same thing are equal also to one another;
2. And if equals be added to equals, the wholes are equal;
3. And if equals be subtracted from equals, the remainders are equal;
4. And things that coincide are equal;
5. And the whole is greater than the part.

Proposition 1

Upon a given finite straight line to construct an equilateral triangle.

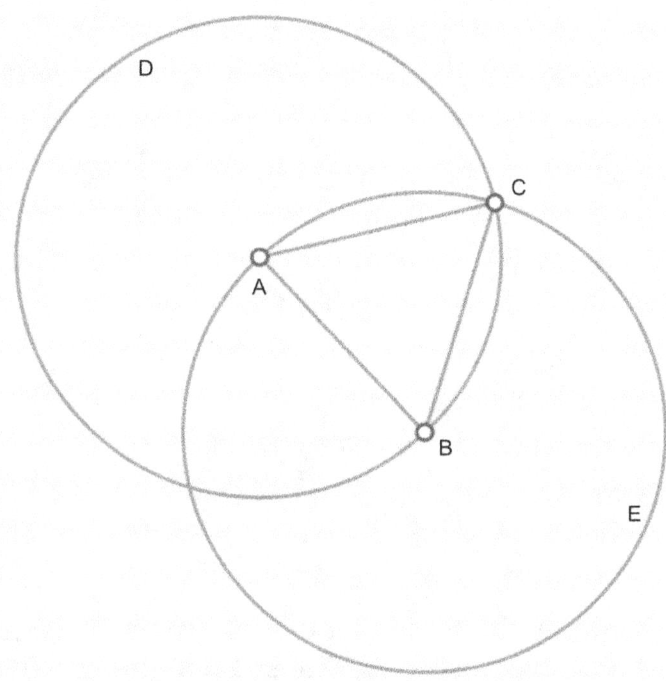

Let there have been given a finite straight line, line AB.
It is required upon the line AB, to construct an equilateral triangle.
Making the center point A, and the distance AB, let there have been described a circle, the circle BCD.
Again, making the center point B, and the distance BA, let there have been described a circle, the circle ACE.
And from the point C, where the circles cut one another, to point A and to point B, let there have been drawn the straight lines CA and CB.
Now, since the point A is the center of the circle BCD, therefore AC = AB.
Again, since the point B is the center of circle ACE, therefore BC = AB.
But it is demonstrated that AC = AB; therefore both of the lines AC and BC are equal to AB.
But things equal to the same thing are equal to one another;
therefore, also AC = BC.
Therefore, the three straight lines, AB, BC, and AC are equal to one another.

Therefore, the triangle ABC is equilateral, and it has been constructed upon the finite straight line AB.

The very thing it was required to do.

Proposition 2

From a given point, to draw a straight line equal to a given straight line.

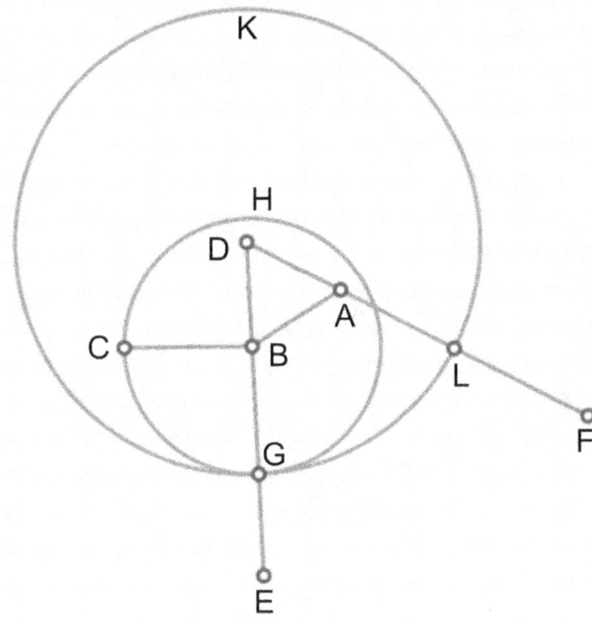

Let there have been given a point, point A, and let there have been given a straight line, line BC.
It is required from the point A to draw a straight line equal to the line BC.
From the point A to the point B, let a straight line have been drawn, line AB.
And upon the line AB let there have been constructed an equilateral triangle, the triangle ABD.
And taking the lines DA and DB let them have been extended in a straight line by AF and BE.
Making the center B, and the distance BC, let there have been described a circle, CGH (where G is the point where this circle cuts BE).
Again, making the center D and the distance DG, let there have been described a circle, GKL (where L is the point where this circle cuts AF).
Now since the point B is the center of circle CGH, BC = BG.
Again since the point D is the center of circle GKL, DL = DG;
of which lines the parts DA and DB are equal.
Therefore the remainders are equal, namely, AL = BG.
But it is demonstrated that BC = BG.
Therefore both of the lines BC and AL are equal to the line BG.
But things equal to the same thing are equal also to one another.
Therefore the line AL is equal to the line BC.

Therefore from a given point A is drawn a straight line AL equal to the given straight line BC.

The very thing it was required to do.

Proposition 3

Two unequal straight lines being given, to cut off from the greater a straight line equal to the less.

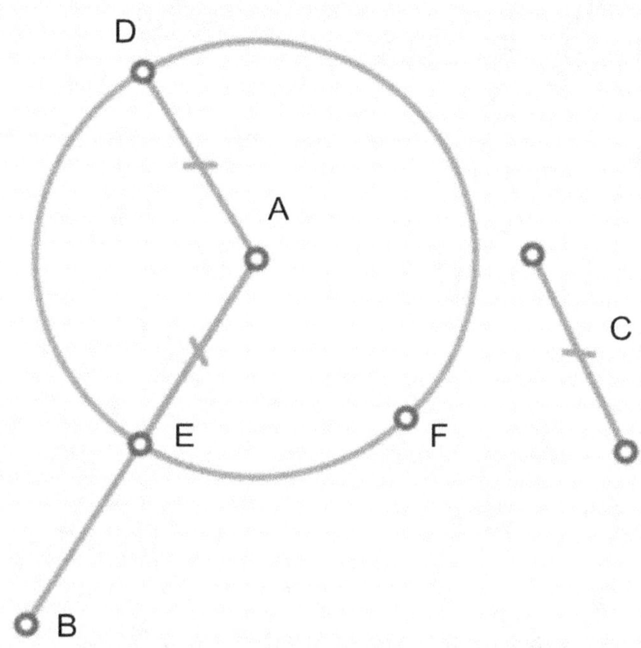

Let there have been given two unequal straight lines, line AB and line C, of which let AB have been the greater.
It is required from the line AB, being the greater, to cut off a straight line equal to the straight line C, the less.
From the point A, let there have been drawn a straight line equal to the line C, the line AD.
And making the center A, and the distance AD, let there have been described a circle, the circle DEF.
Now, since the point A is the center of the circle DEF, AE = AD.
But the line C is equal to the line AD.
Therefore, both of the lines AE and C are equal to the line AD.
Therefore AE = C.

Therefore, two unequal straight lines being given, namely AB and C, there is cut off from AB the greater a straight line AE equal to the less, namely C.

The very thing it was required to do.

Proposition 4

If two triangles have two sides respectively equal to two sides, and the angles comprehended by the equal sides equal; then also the base of the one shall be equal to the base of the other, and the one triangle shall be equal to the other, and the other angles remaining shall be respectively equal to one another, namely those that are subtended by equal sides.

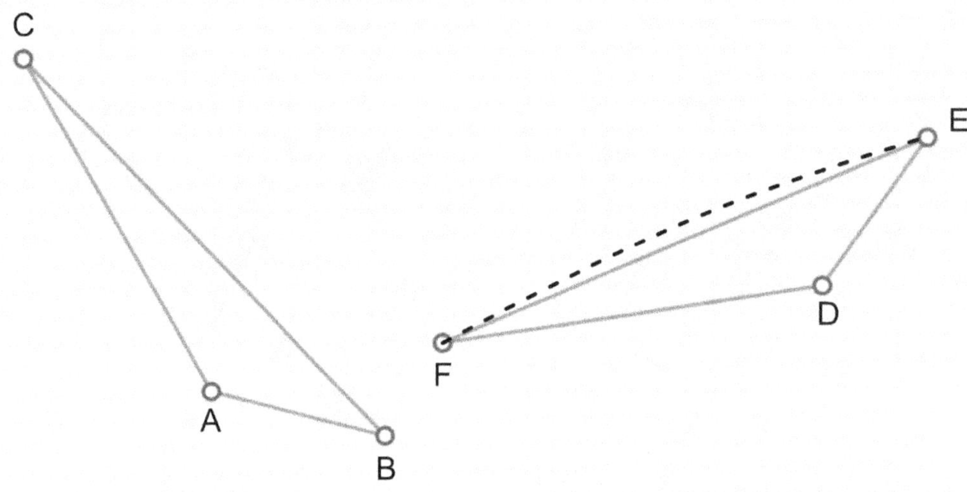

Let there have been given two triangles, ABC and DEF, having two sides of the one, AB and AC, respectively equal to two sides of the other, DE and DF; that is, AB = DE and AC = DF; having also the angle BAC equal to the angle EDF.
I say that the base, BC is equal to the base, EF; and the triangle ABC = the triangle DEF; and the other angles remaining are respectively equal to the other angles remaining, namely those that are subtended by equal sides; that is, angle ABC = angle DEF and angle ACB = angle DFE.
Now if triangle ABC be applied to triangle DEF, and the point A be put on the point D, and the line AB be put on the line DE, then the point B shall also coincide with point E because AB = DE. And the line AB coinciding with the line DE, the straight line AC shall coincide with the straight line DF because the angle BAC = angle EDF. Therefore also the point C shall coincide with the point F because AC = DF.
Again, the point C coincides with point F, and the point B coincides with point E; therefore, the base BC shall coincide with the base EF.
For, if B coincides with E, and C coincides with F, and the base BC does not coincide with EF then two straight lines will enclose a space, which is impossible.
Therefore, the base BC coincides with the base EF; and therefore is equal to it.
Therefore the whole triangle ABC coincides with the whole triangle DEF, and therefore is equal to it.
And the other remaining angles coincide respectively with the other remaining angles, and therefore are equal to one another; that is, angle ABC = DEF and ACB = DFE.

If therefore two triangles have two sides respectively equal to two sides, and the angles comprehended by the equal sides equal; then also the base of the one shall be equal to the base of the other, and the one triangle shall be equal to the other, and the other angles remaining shall be respectively equal to one another, namely those that are subtended by equal sides.

The very thing it was required to demonstrate.

Proposition 5

An isosceles triangle has its angles at the base equal to one another. And those equal sides being extended, the angles which are under the base are also equal to one another.

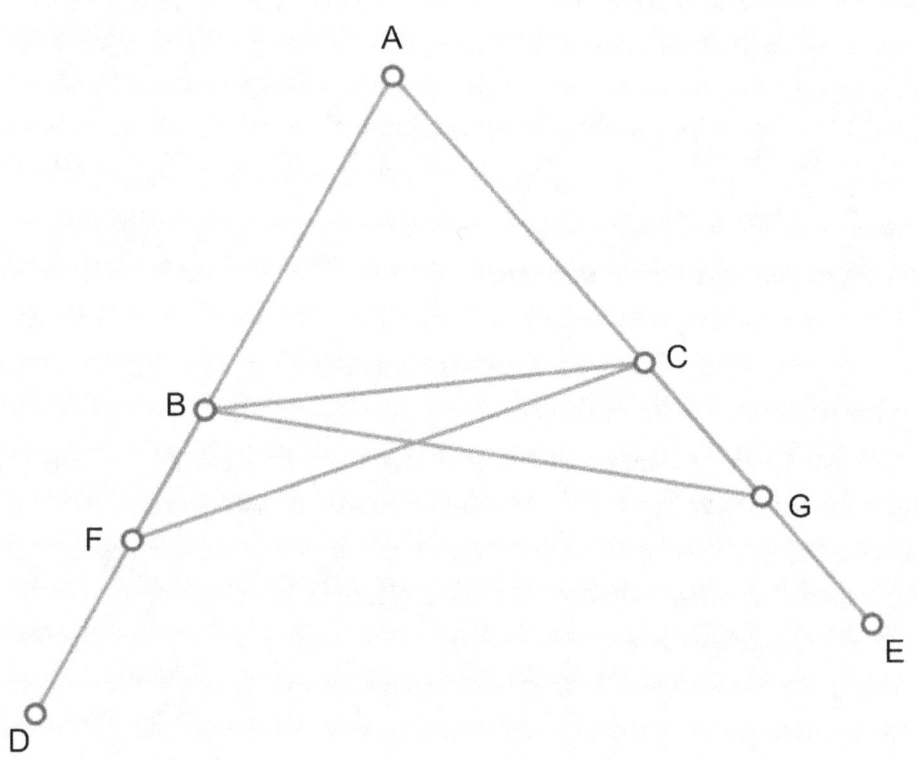

Let there have been given an isosceles triangle, triangle ABC, having side AB = AC. And let the lines AB and AC have been extended in a straight line producing BD and CE.
I say that angle ABC = ACB; and angle CBD = angle BCE.
On the line BD, let there have been taken any point whatsoever, point F.
And from the greater, AE, let there have been cut off a line equal to AF, the line AG.
And let there have been drawn CF and BG.

Now, since AF = AG, and AB = AC, the two lines AF and AB are respectively equal to the two lines AG and AC;
and they contain a common angle, namely the angle contained under FAG;
therefore, base CF is equal to the base BG, and the triangle ACF = triangle ABG, and the other angles remaining shall be respectively equal to one another, namely those that are subtended by equal sides; that is, angle ACF = angle ABG and angle AFC = angle AGB.
And since the whole AF = the whole AG, of which the parts AB = AC;
therefore the remainders are equal: BF = CG.
Now therefore, two lines BF and CF are respectively equal to the two lines CG and BG; and angle BFC = angle CGB, and they have one base common to both of them, namely BC;
therefore, triangle CBF = triangle BCG, and the other angles remaining shall be respectively equal to one another, namely those that are subtended by equal sides;
therefore, angle FBC = angle GCB and angle BCF = angle CBG.
Now since the whole angle ABG = the whole angle ACF, of which the parts angle CBG = angle BCF;
therefore the remainders are equal: angle ABC = angle ACB, and they are the angles at the base of the triangle ABC. And it is demonstrated that angle FBC = angle GCB; and they are the angles under the base.

Therefore an isosceles triangle has its angles at the base equal to one another. And those equal sides being extended, the angles which are under the base are also equal to one another.

The very thing it was required to demonstrate.

Proposition 6

If in a triangle two angles are equal to one another, the sides of the triangle, which subtend the equal angles, will also be equal to one another.

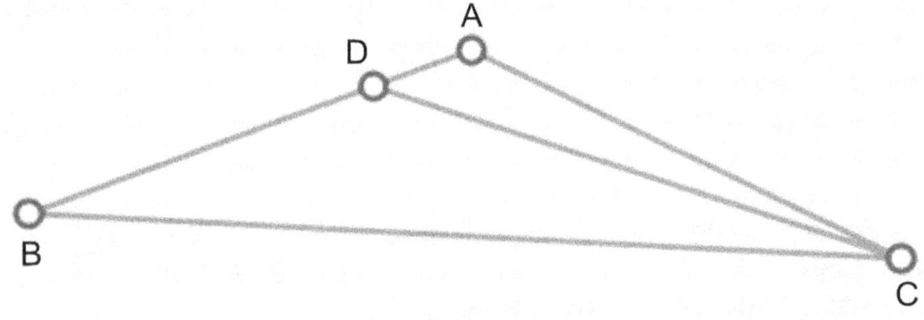

Let a triangle have been given, triangle ABC, having the angle ABC equal to the angle ACB.
I say that the side AB is also equal to the side AC.
For if the side AB is not equal to the side AC, then one of them is greater.
Let AB be the greater.
From AB the greater, let there have been cut off a line equal to AC the less, namely line DB.
And let the line DC have been drawn.
Now, since the side DB equals the side AC, and line BC is common to both;
therefore, the two sides DB and BC are respectively equal to the two sides AC and BC.
And the angle DBC is given equal to the angle ACB.
Therefore, the base DC = base AB, and triangle DBC = triangle ACB: the less equal to the greater,
which is absurd.
Therefore the side AB is not unequal to the side AC.
Therefore they are equal.

If therefore in a triangle two angles are equal to one another, the sides of the triangle, which subtend the equal angles, will also be equal to one another.

The very thing it was required to demonstrate.

Proposition 7

If from the same ends of one line, two straight lines be drawn to any point, there can not be drawn from the same ends on the same side two other lines respectively equal to the first two lines at any other point.

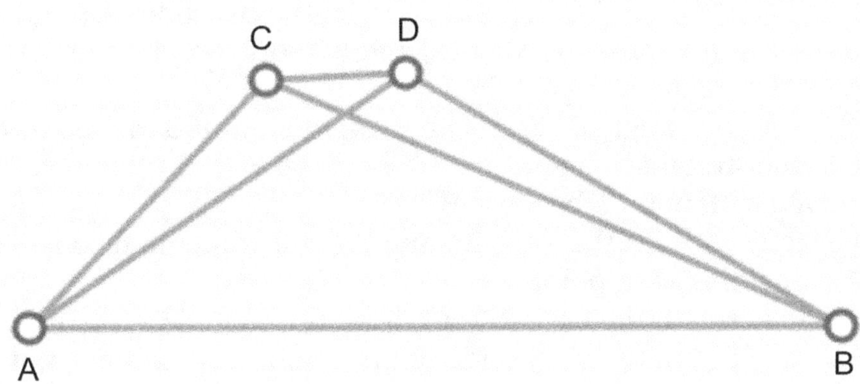

Or if it be possible, then from the same ends of one straight line, namely AB, from the ends, A and B, let there have been drawn two straight lines, namely AC and BC, to the same point, C. And from the same ends of the line AB and on the same side, let there have been drawn two other straight lines, AD and BD respectively equal to the lines AC and BC, meeting at another point, point D.

And let the point C have been joined by a straight line to point D.
Now, since AC = AD, angle ACD = angle ADC.
Therefore angle ACD is less than angle BDC.
Therefore angle BCD is much less than angle BDC.
Again, since BC = BD, the angle BCD = angle BDC; but it is also demonstrated that the first is much less than the other; which is absurd.

If therefore, from the ends of one line, two straight lines be drawn to any point, there can not be drawn from the same ends on the same side two other lines respectively equal to the first two lines at any other point.

The very thing it was required to demonstrate.

Proposition 8

If two triangles have two sides respectively equal to two sides, and also have the base of one equal to the base of the other, they shall have also the angles contained by the respectively equal straight lines equal to one another.

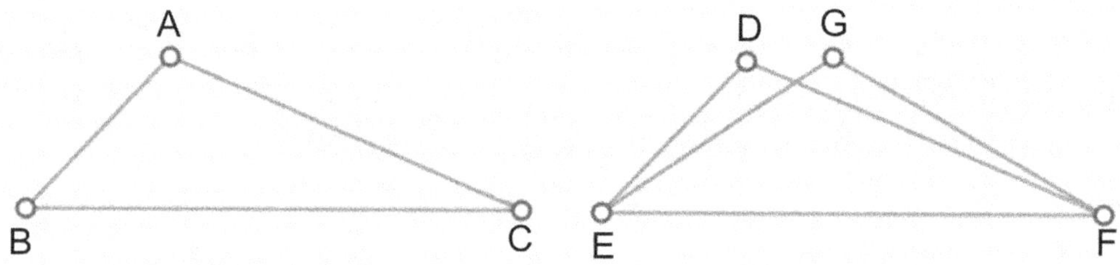

Let two triangles have been given, ABC and DEF; and let the two sides AB and BC be respectively equal to the two sides DE and EF; that is, AB = DE and BC = EF; and let the base AC = the base DF. I say that angle BAC is also equal to angle EDF.
For applying triangle ABC to triangle DEF, and applying the point B to the point E, and applying the straight line BC to the straight line EF;
the point C will coincide with point F because BC = EF.
And BC coinciding exactly with EF, the lines BA and CA will coincide respectively with the sides ED and FD (point A coinciding with point D).
For if the base BC coincides with the base EF, but the sides BA and CA do not coincide respectively with ED and FD, but fall to the side (point A falling at point G, not point D) forming lines EG and FG;
then from the same ends of one line, two straight lines shall be drawn to a point, and from the same ends on the same side shall be drawn two other lines respectively equal to the first two lines at another point:
but that is impossible.
Therefore, the base BC coinciding with the base EF, the sides also BA and AC coincide respectively with the sides ED and DF.
Therefore the angle BAC coincides with the angle EDF,
and therefore is equal to it.

If therefore two triangles have two sides respectively equal to two sides, and also have the base of one equal to the base of the other, they shall have also the angles contained by the respectively equal straight lines equal to one another.

The very thing it was required to demonstrate.

Proposition 9

To bisect a given rectilineal angle.

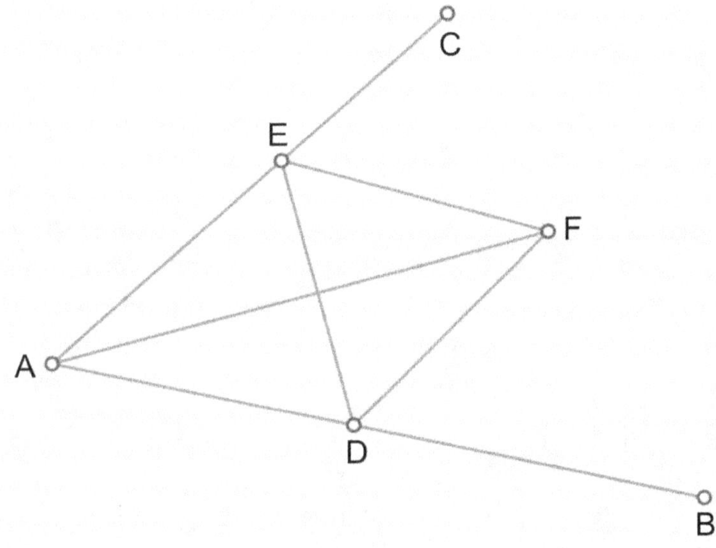

Let a rectilineal angle have been given, angle BAC.
It is required to divide angle BAC into two equal parts.
On the line AB, let there have been taken any point whatsoever, point D.
And from the line AC let there have been cut off a line equal to AD, the line AE.
And let the line DE have been drawn.
And upon the line DE let an equilateral triangle have been described, the triangle EDF.
And let the line AF have been drawn.
I say that the angle BAC is bisected by the line AF.
For, since AD = AE, and AF is common to both,
these two AD and AF are respectively equal to the other two AE and AF.
And the base DF = the base EF.
Therefore, the angle DAF = the angle FAE.

Therefore the given rectilineal angle, BAC, is bisected by the straight line AF.

The very thing it was required to do.

Proposition 10

To bisect a given finite straight line.

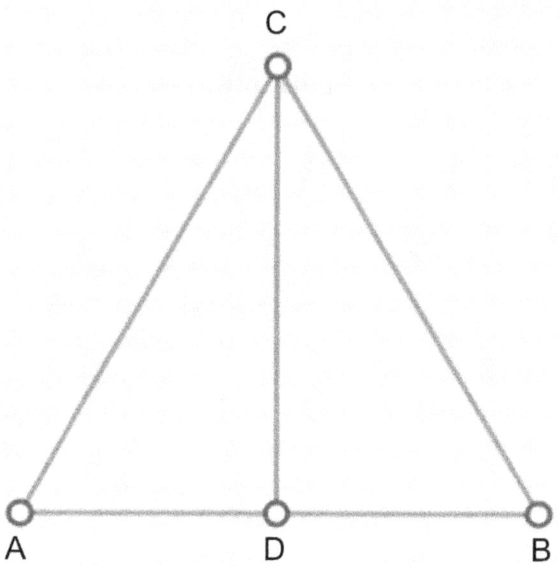

Let a finite straight line be given, the line AB.
It is required to divide the line AB into two equal parts.
Upon AB let an equilateral triangle have been described, triangle ABC.
And let the angle ACB have been bisected by the line CD.
I say that the straight line AB is bisected at the point D.
For since AC = CB, and CD is common to both,
the two lines AC and CD are respectively equal to the two lines BC and CD.
And the angle ACD = angle BCD.
Therefore, base AD = base DB.

Therefore, the given finite straight line AB is bisected at point D.

The very thing it was required to do.

Upon a given straight line, at a given point on the same line, to raise up a perpendicular line.

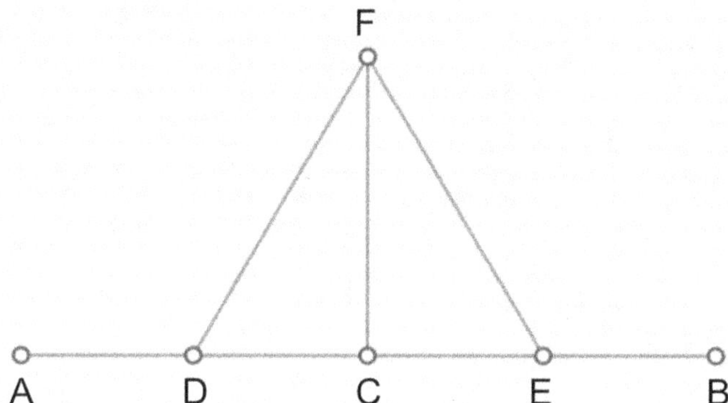

Let a straight line have been given, AB; and let a point on it have been given, C.
It is required from the point C to raise up a line perpendicular to AB.
On line AC, let there have been taken any point whatsoever, point D.
And let the line CE (cut off from CB or from the extension of CB) have been made equal to CD.
And upon DE, let an equilateral triangle have been described, triangle DEF.
And let the line CF have been drawn.
I say that the line CF has been raised up perpendicular to the given line AB, at the given point C.
For, since DC = CE, and the line CF is common to both,
the two DC and CF are respectively equal to the two CE and CF.
And base DF = base EF.
Therefore, angle DCF = angle ECF.
And they are adjacent angles.
But when a straight line standing upon a straight line makes the adjacent angles equal,
then either of those angles is a right angle. And the straight line that stands erect is called a perpendicular to that upon which it stands
Therefore, the angles DCF and ECF are right angles.

Therefore, upon the straight line AB, at the given point on it (point C), there is raised up a perpendicular line, CF.

The very thing it was required to do.

Proposition 12

Unto a given infinite straight line, from a given point not on it, to draw a perpendicular line.

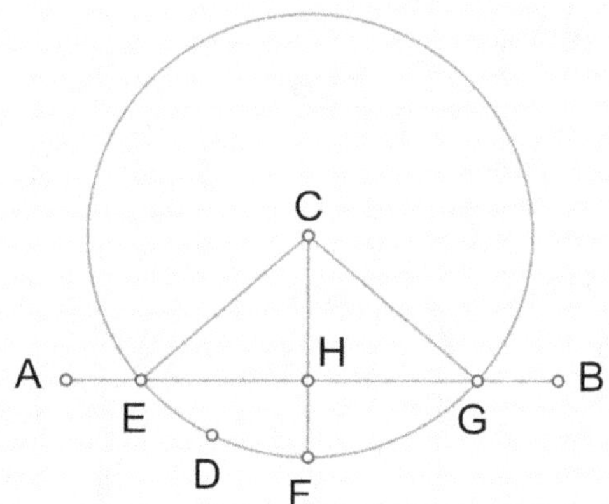

Let a straight line have been given, AB; and let a point not on AB be given, C.
It is required from the given point C, to draw unto the given infinite straight line AB, a perpendicular line.
On the other side of line AB (namely the side where point C is not), let there have been taken a point at random, D.
And making C the center and CD the distance, let there have been described a circle EFG, which cuts the line AB at two points, G and E.
And let the line GE have been bisected at a point, point H.
And let the lines CG, CE, and CH have been drawn.
I say that unto the given infinite straight line AB, and from the given point not on it C, is drawn a perpendicular line, CH.
For since GH = EH, and CH is common to them both;
the two sides GH and CH are respectively equal to the two sides EH and CH.
And the base CG = the base CE.
Therefore angle CHG = angle CHE;
and they are adjacent angles.
But when a straight line standing upon a straight line makes the adjacent angles equal,
then either of those angles is a right angle. And the straight line that stands erect is called a perpendicular to that upon which it stands.

Therefore unto a given infinite straight line, AB; from a given point, C, not on line AB; is drawn a perpendicular line, CH.

The very thing it was required to do.

When a straight line stands upon a straight line making any angles, those angles will be either two right angles or equal to two right angles.

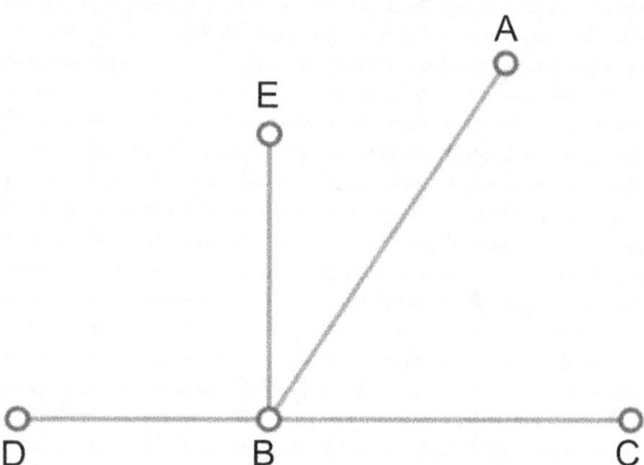

Let a straight line, AB, standing upon another straight line, CD, make the angles CBA and ABD.
I say that the angles CBA and ABD are either two right angles or else equal to two right angles.
If angle CBA be equal to angle ABD, then they are two right angles.
But if not, let there have been raised upon the straight line CD, and from the given point on it, point B, a perpendicular line, EB.
Therefore, the angles CBE and EBD are right angles.
Now since the angle CBE is equal to the two angles ABC and ABE, add the angle EBD in common to both; therefore, the two angles CBE and DBE are equal to the three angles ABC, ABE, and DBE.
Again, since the angle DBA is equal to the two angles DBE and ABE, add the angle ABC in common to both;
therefore, the two angles DBA and ABC are equal to the three angles DBE, ABE, and ABC.
And it is demonstrated that the angles CBE and EBD are equal to the same three angles.
But things equal to the same thing are equal also to one another.
Therefore, the two angles, CBE and DBE, are equal to the two angles, DBA and ABC.
But the angles CBE and DBE are two right angles;
therefore, the two angles, DBA and ABC, are equal to two right angles.

Therefore, when a straight line stands upon a straight line making any angles, those angles will be either two right angles or equal to two right angles.

The very thing it was required to demonstrate.

If unto a straight line, and at one point on the same, be drawn two straight lines, not both on the same side, forming adjacent angles equal to two right angles; those two straight lines will lie in one straight line.

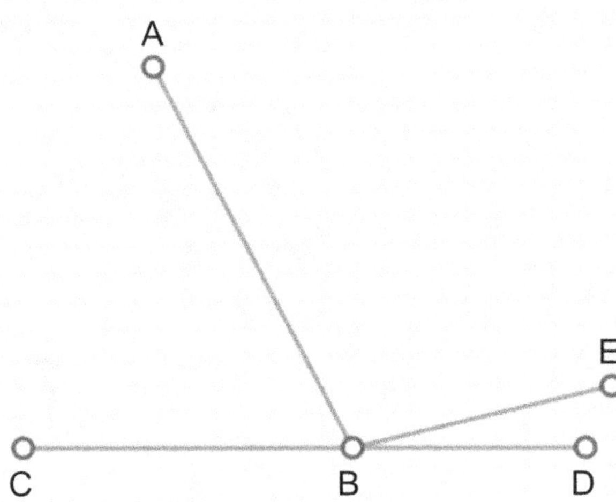

Unto the straight line AB, and at one point on it, B, let there have been drawn two straight lines, BC and BD, on opposite sides, making the adjacent angles, ABC and ABD, equal to two right angles.
I say that lines BD and BC lie in one straight line.
For if BD and BC do not lie in one straight line, let there have been drawn a straight line , BE, continuously in a straight line with CB.
Now, since the straight line AB stands upon the straight line CBE;
therefore, the two angles ABC and ABE are equal to two right angles.
But the two angles ABC and ABD are equal to two right angles;
therefore, the two angles ABC and ABE are equal to the two angles ABC and ABD.
Subtract the angle ABC which is common to both.
Therefore the remainder, angle ABE, is equal to the remainder, angle ABD; the less equal to the greater; which is impossible.
Therefore, the line BE is not so drawn continuously from CB that they lie in a single straight line.
In like manner, we can demonstrate that no other line, except BD, can be so drawn.
Therefore, the lines CB and BD lie in one straight line.

If therefore, unto a straight line, and at one point on the same, be drawn two straight lines, not both on the same side, forming adjacent angles equal to two right angles; those two straight lines will lie in one straight line.

The very thing it was required to demonstrate.

If two straight lines cut one another, the vertical angles will be equal to one another.

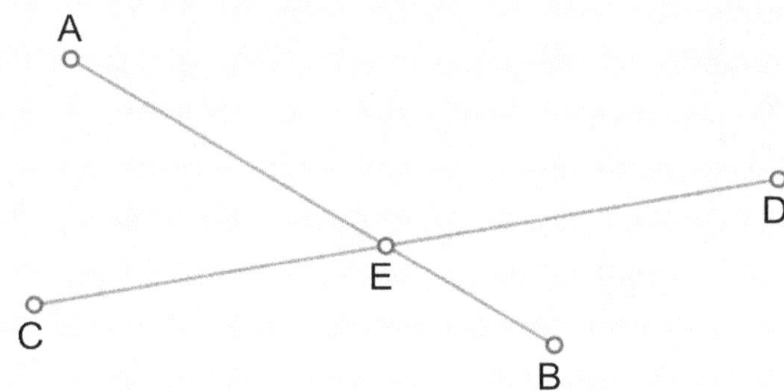

Let the two straight lines, AB and CD, cut one another at a point, E.
I say that angle AEC = angle BED and angle BEC = angle AED.
For, since the straight line AE stands upon the straight line DC, making the angles AEC and AED;
therefore, the two angles, AEC and AED, are equal to two right angles.
Again, since the straight line DE stands upon the straight line AB, making the angles AED and BED;
therefore, the two angles, AED and BED, are equal to two right angles.
And it is demonstrated that the two angles AEC and AED are also equal to two right angles.
Therefore, the two angles, AEC and AED, are equal to the two angles, AED and BED.
Subtract the angle AED which is common to both.
Therefore, the remainder, angle AEC, is equal to the remainder, angle BED.
And in like manner it may be demonstrated that angle BEC = AED.

Therefore, if two straight lines cut one another, the vertical angles will be equal to one another.

The very thing it was required to demonstrate.

In every triangle, if one side is extended, the exterior angle will be greater than either of the two interior and opposite angles.

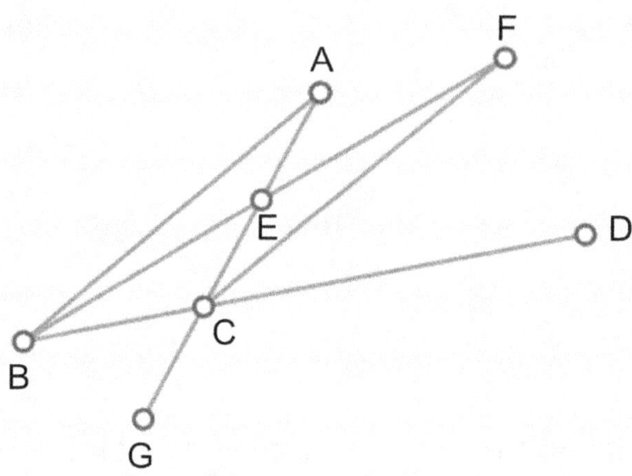

Let a triangle have been given, triangle ABC.
And let one of the sides, side BC, have been extended to a point D.
I say that the exterior angle ACD is greater than either of the two interior and opposite angles, that is angle ABC or angle BAC.
Let the line AC have been bisected at point E.
And let the line BE have been drawn.
And let the line BE have been extended to F,
and cut off so that EF = BE.
And let the line FC have been drawn.
And let the line AC have been extended to G.
Now, since AE = CE, and BE = EF; the two sides, AE and EB, are respectively equal to the two sides CE and EF.
And angle AEB = angle FEC, for they are vertical angles.
Therefore, the base AB = the base FC, and triangle ABE = triangle FEC, and the remaining angles are respectively equal, namely those that are subtended by equal sides.
Therefore angle BAE = angle ECF.
But the angle ECD is greater than the angle ECF.
Therefore the angle ACD is greater than the angle BAC.
In like manner also, if the line BC be bisected, it can be demonstrated that the angle BCG, that is, the angle ACD, is greater than the angle ABC.

Therefore, in every triangle, if one side is extended, the exterior angle will be greater than either of the two interior and opposite angles.

The very thing it was required to demonstrate.

In every triangle, any two angles taken together are less than two right angles.

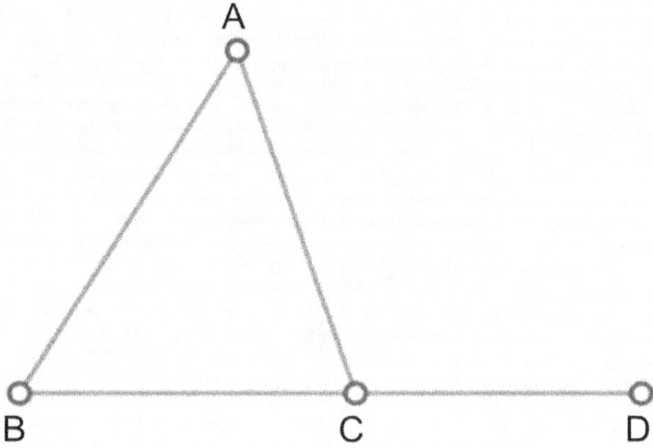

Let there have been a triangle, ABC.
I say that any two angles of the said triangle taken together, are less than two right angles.
Let the line BC have been extended to D.
Then, since the exterior angle of triangle ABC, namely, angle ACD, is greater than the interior and opposite angle ABC, add the angle ACB in common to both.
Therefore, the two angles, ACD and ACB, are greater than the two angles, ABC and ACB.
But the two angles, ACD and ACB, are equal to two right angles.
Therefore, the two angles, ABC and ACB, are less than two right angles.
In like manner, we can also demonstrate that the two angles, BAC and ACB, are less than two right angles, and also that the two angles, CAB and ABC, are less than two right angles.

Therefore, in every triangle, any two angles taken together are less than two right angles.

The very thing it was required to demonstrate.

In every triangle, to the greater side is subtended the greater angle.

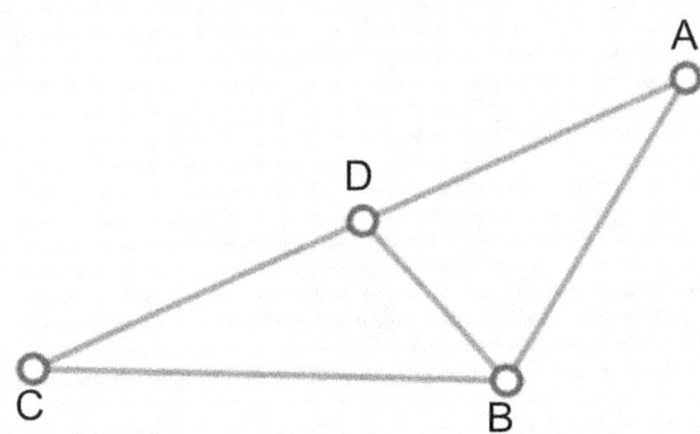

Let there have been a triangle, ABC; having one side, AC, greater than another side, AB.
I say that the angle ABC is greater than the angle BCA.
For since AC is greater than AB, let there have been cut off a line equal to AB, the line AD.
And let the line BD have been drawn.
Now, the exterior angle, ADB, is greater than the interior and opposite angle, DCB.
But angle ADB = angle ABD for the side AB = the side AD.
Therefore, angle ABD is greater than angle ACB.
Therefore, angle ABC is much greater than angle ACB.

Therefore, in every triangle, to the greater side is subtended the greater angle.

The very thing it was required to demonstrate.

In every triangle, under the greater angle is subtended the greater side.

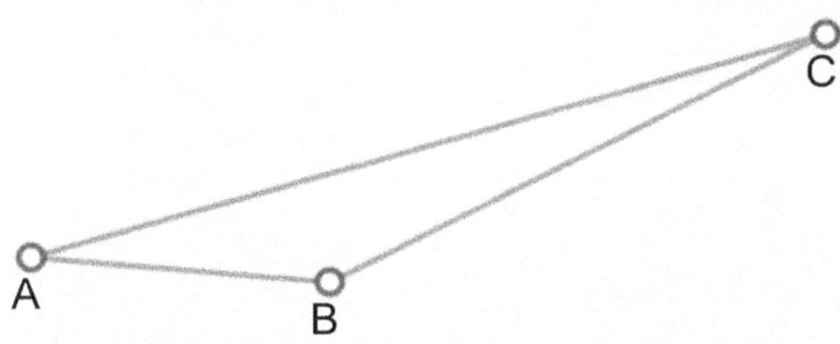

Let there have been a triangle, ABC; having one angle, ABC, greater than another, BCA.
I say that the side AC is greater than the side AB.
For if not, then the side AC is either equal to the side AB or it is less than it.
The side AC is not equal to the side AB, for then angle ABC should be equal to angle ACB; but it is not.
Therefore, the side AC is not equal to the side AB.
And the side AC is not less than the side AB, for then angle ABC should be less than the angle ACB, but it is not.
Therefore, the side AC is not less than side AB.
Therefore, the side AC is greater than the side AB.

Therefore, in every triangle, under the greater angle is subtended the greater side.

The very thing it was required to demonstrate.

In every triangle, any two sides taken together are greater than the remaining side.

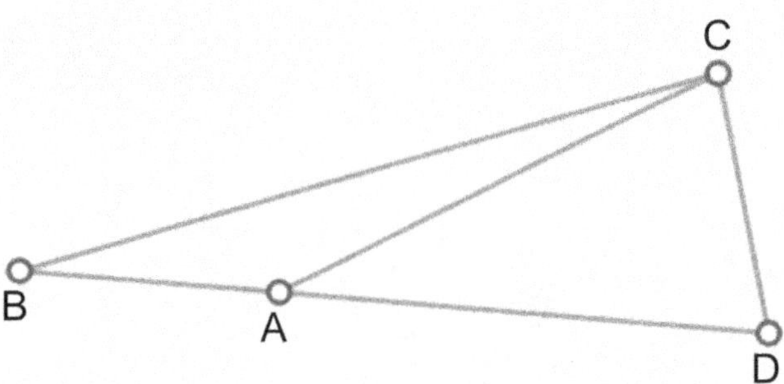

Let there have been given a triangle, ABC.
I say that any two sides of the triangle, ABC, taken together are greater than the remaining side;
that is, AB and AC together are greater than BC; AB and BC together are greater than AC; and,
AC and BC together are greater than AB.
Let the line BA have been extended to point D,
letting AD have been cut off equal to AC.
And let the line CD have been drawn.
Now, since the line DA = the line AC,
the angle ADC = the angle ACD.
But the angle BCD is greater than the angle ACD,
therefore, the angle BCD is greater than the angle ADC.
And since DCB is a triangle, having the angle BCD greater than the angle ADC,
but under the greater angle is subtended the greater side;
therefore, DB is greater than BC.
But AD = AC.
Therefore, the line DB is equal to the lines AB and AC together.
Therefore, the lines AB and AC together are greater than the line BC.
And in like manner we could demonstrate that the sides AB and BC are greater than AC, and
that the sides BC and CA are greater than the side AB.

Therefore, in every triangle, any two sides taken together are greater than the remaining side.

The very thing it was required to demonstrate.

Proposition 21

If two straight lines be drawn from the ends of one of the sides of a triangle to any point within the triangle, those two straight lines drawn shall be less than the two other sides of the triangle but will contain the greater angle.

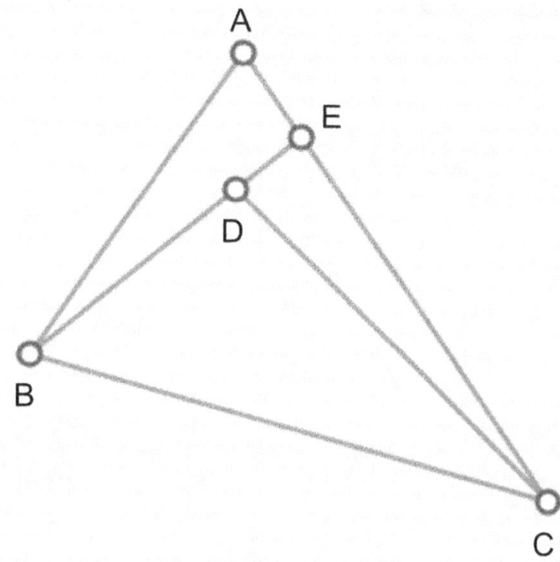

Let there have been a triangle, ABC; and from the ends of one side, BC, from the points B and C, let there have been drawn two straight lines, BD and CD, to a point within the triangle, point D.
I say that the lines BD and CD are less than the other sides of the triangle, namely, the sides AB and AC; and that the angle which they contain, BDC, is greater than the angle BAC.
Let the line BD have been extended to the point E.
And since in every triangle any two sides are greater than the remaining side,
therefore, the two sides of the triangle ABE, namely AB and AE, are greater than the side EB.
Add the line EC to both.
Therefore the lines BA and AC together are greater than the lines BE and EC together.
Again, since in the triangle CED, the two sides CE and ED together are greater than the side DC, add the line DB to both,
therefore, the lines CE and EB together are greater than the lines CD and DB together.
But it is proved that BA and AC together are greater than BE and EC together.
Therefore, the lines BA and AC together are much greater than CD and DB together.
Again, since in every triangle, the exterior angle is greater than the opposite interior angles, the exterior angle of triangle CDE, namely BDC is greater than the angle CED.
Also, the exterior angle of the triangle ABE, namely CEB, is greater than the angle BAC.
Therefore the angle BDC is much greater than the angle BAC.

Therefore if two straight lines be drawn from the ends of one of the sides of a triangle to any point within the triangle, those two straight lines drawn shall be less than the two other sides of the triangle but will contain the greater angle.

The very thing it was required to demonstrate.

Of three straight lines, that are equal to three given straight lines, to construct a triangle. But it is necessary that any two of the lines, taken together in any order, be greater than the third, for in that every triangle, any two sides taken together are greater than the remaining side.

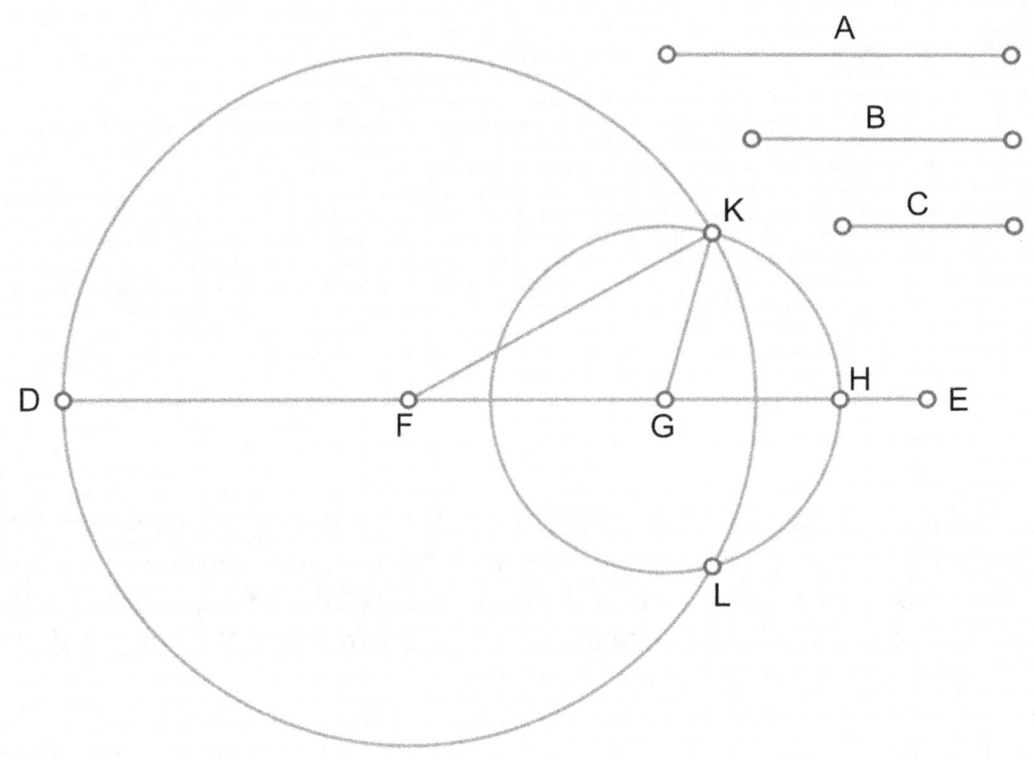

Let there have been given three straight lines, lines A, B, and C;
of which any two taken together are greater than the third;
that is,
A and B together are greater than C,
A and C together are greater than B,
and B and C together are greater than A.
It is required that of three straight lines equal to the straight lines A, B, and C to construct a triangle.
Let there have been a straight line having an end point at D and being indefinitely continued to point E.
And let there have been cut off equal to A the line DF, and equal to B the line FG, and equal to C the line GH.
And making the center F, and the distance DF, let there have been described a circle, the circle DKL.

Again making the center G, and the distance GH, let there have been described a circle, the circle HKL.
And let there have been drawn the straight lines FK and GK.
I say that, of three straight lines equal to A, B, and C, is constructed a triangle FGK.
For since the point F is the center of circle DKL, FD = FK.
But line A = FD; therefore, FK = line A.
Again since the point G is the center of circle HKL, GK = GH.
But line C = GH; therefore, GK = line C.
And by construction FG = line B.
Therefore, the three straight lines FK, GK, and FG are equal to the three straight lines, A, B, and C.

Therefore of three straight lines FK, GK and FG, that are equal to three given straight lines A, B, and C, is constructed a triangle FGK.

The very thing it was required to do.

On a given straight line, and at a given point on it, to construct a rectilineal angle equal to a given rectilineal angle.

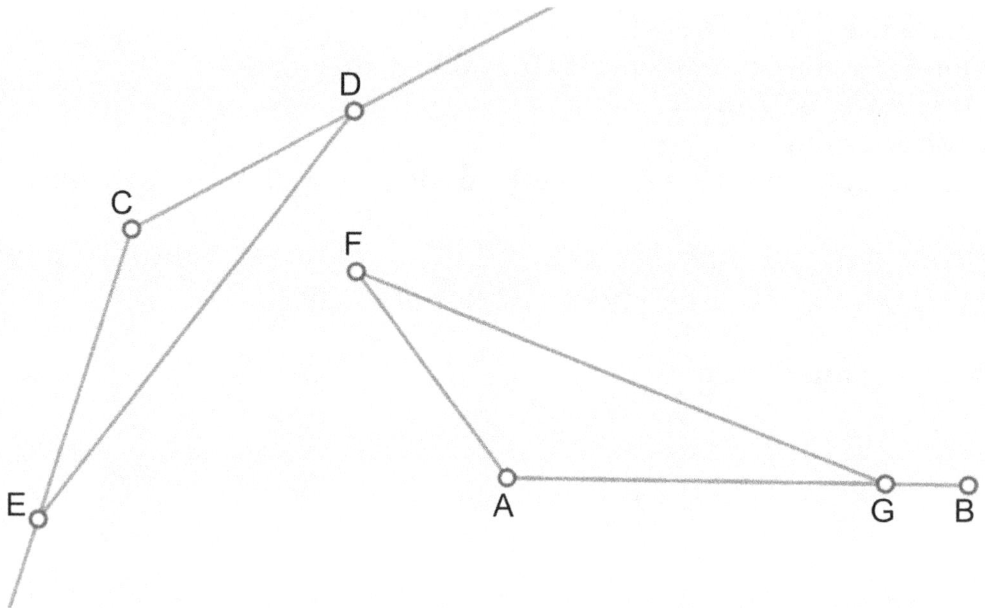

Let there have been given a straight line, AB; and let there have been given a point on it, point A. And also let there have been given a rectilineal angle, angle DCE.
It is required upon the given straight line AB, at the given point A, to construct a rectilineal angle equal to the given rectilineal angle DCE.
On the lines DC and EC let there have been taken at random the points D and E.
Let there have been drawn the line ED.
And of three straight lines equal to lines CE, ED, and CD, let there have been constructed a triangle AFG such that CD = AF, CE = AG, and DE = FG.
Now since the two lines, DC and CE are respectively equal to the two lines AF and AG, and the base DE = the base FG, angle DCE = angle FAG.

Therefore, on a given straight line, AB, and at a given point on it, A, is constructed a rectilineal angle, FAG, equal to a given rectilineal angle, DCE.

The very thing it was required to do.

Proposition 24

If two triangles have two sides of the one respectively equal to two sides of the other, and if the angle contained by the equal sides of the one be greater than the angle contained by the equal sides of the other, the base of the one shall be greater than the base of the other.

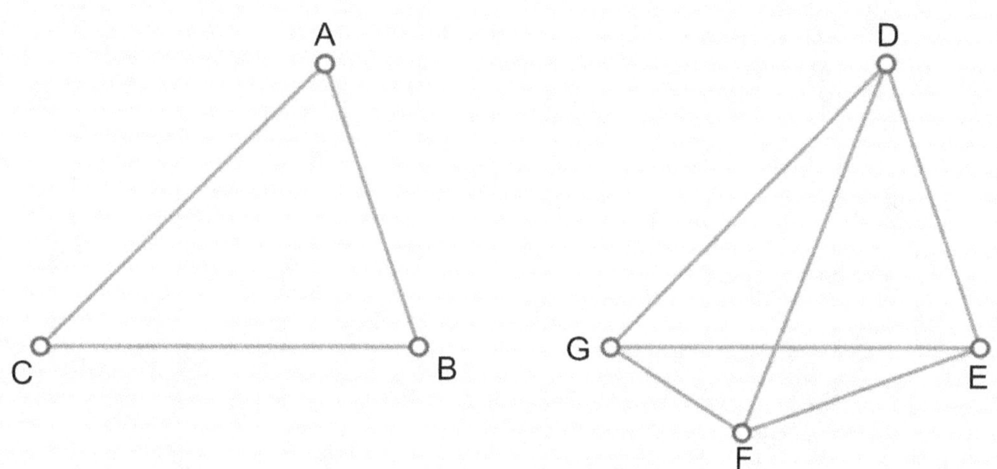

Let there have been given two triangles, ABC and DEF, having the two sides of the one, AB and AC, equal respectively to the two sides of the other, DE and DF, such that AB = DE and AC = DF.
And let the angle BAC have been greater than the angle EDF.
I say that the base BC is greater than the base EF.
For, since the angle BAC is greater than the angle EDF, upon the straight line DE, at point D, let there have been constructed an angle equal to BAC, the angle EDG, letting the line DG have been cut off equal to either AC or DF.
And let there have been drawn the lines FG and EG.
Now since AB = DE, and AC = DG, and angle BAC = angle EDG;
therefore, the base BC = base EG.
Again, since DG = DF, angle DGF = angle DFG.
Therefore, angle DFG is greater than angle EGF.
Therefore, angle EFG is much greater than EGF.
And since EFG is a triangle, having angle EFG greater than angle EGF, and under the greater angle is subtended the greater side;
therefore, the side EG is greater than the side EF.
But the side EG = side BC, therefore the side BC is greater than the side EF.

If therefore, two triangles have two sides of the one respectively equal to two sides of the other, and if the angle contained by the equal sides of the one be greater than the angle contained by the equal sides of the other, the base of the one shall be greater than the base of the other.

The very thing it was required to demonstrate.

If two triangles have two sides of the one respectively equal to two sides of the
other, and if the base of the one be greater than the base of the other, then the
angle contained by the equal sides of the one shall be greater than the angle
contained by the equal sides of the other.

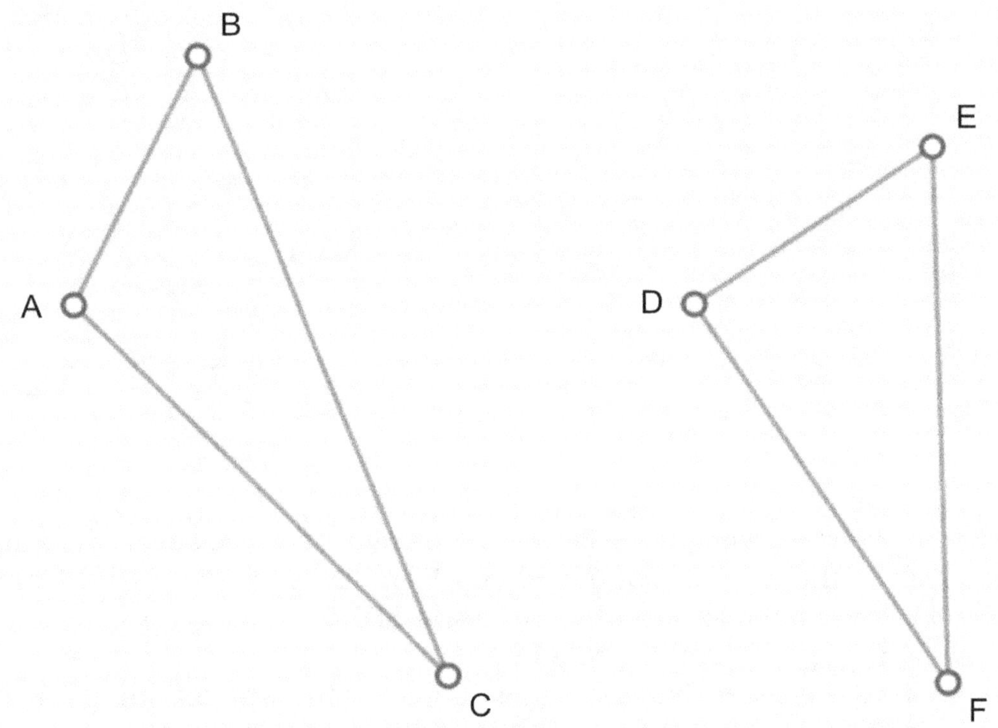

Let there have been given two triangles, ABC and DEF, having the two sides of the one, AB and
AC, equal respectively to the two sides of the other, DE and DF, such that AB = DE and AC = DF.
And let the base BC have been greater than the base EF.
I say that angle BAC is greater than angle EDF.
For if not, then it is either equal to it or less than it.
But the angle BAC is not equal to the angle EDF,
for if it were equal, the base BC would be equal to base EF. But it is not.
Neither is angle BAC less than angle EDF,
for then would the base BC be less than the base EF. But it is not.
Therefore angle BAC is not less than EDF.
And it is proved that it is not equal to it.
Therefore, angle BAC is greater than angle EDF.

If therefore two triangles have two sides of the one respectively equal to two sides
of the other, and if the base of the one be greater than the base of the other, then
the angle contained by the equal sides of the one shall be greater than the angle
contained by the equal sides of the other.

The very thing it was required to demonstrate.

If two triangles have two angles of the one respectively equal to two angles of the
other, and have also one side of the one equal to one side of the other, either the
side that lies between the two equal angles, or the side that is subtended by one of
the equal angles, the other sides of the one shall be respectively equal to the other
sides of the other, and the other angle of the one shall be equal to the other angle
of the other.

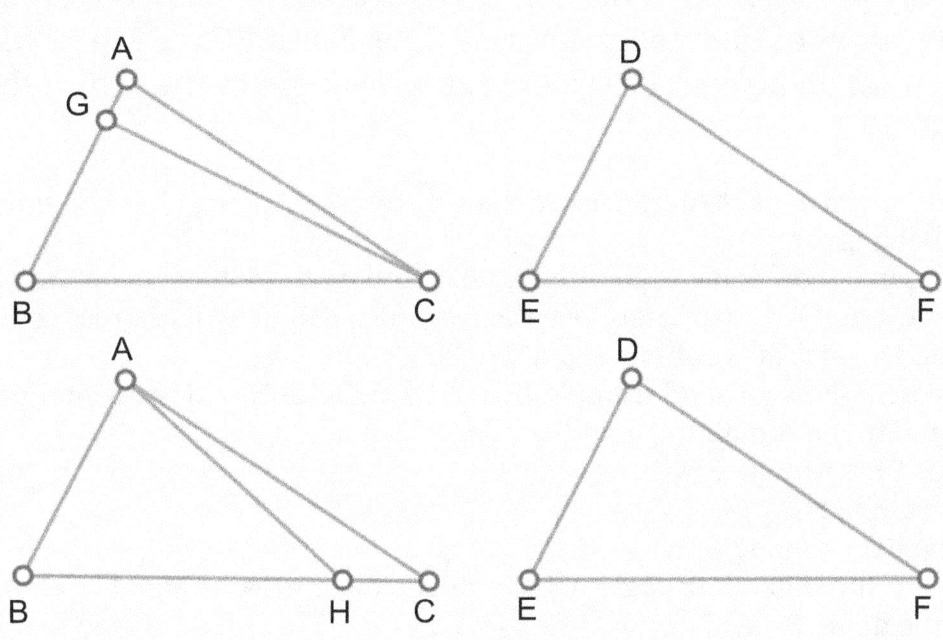

Let there have been given two triangles, ABC and DEF, having two angles of the one, ABC and
BCA respectively equal to two angles of the other, DEF and EFD; that is, angle ABC = angle DEF;
and angle BCA = angle EFD, and one side equal to one side of the other, first, the side that lies
between the equal angles; that is, side BC = side EF.
I say that the other sides of the one shall be respectively equal to the corresponding sides of the
other; that is, AB = DE and AC = DF; and the other angle of the one shall be equal to the other
angle of the other; that is, angle CAB = angle EDF.
For if the side AB is not equal to the side DE, one of them is greater. Let the side AB have been
greater.
Let there have been cut off equal to DE the line GB.
Let there have been drawn the line GC.
Now since GB=DE, and BC = EF, the two lines GB and BC are respectively equal to the two lines
DE and EF, and the included angle GBC = angle DEF.
Therefore, the base GC = base DF and the triangle GCB = triangle DFE, and the remaining angles
of the one are respectively equal to the remaining angles of the other.
Therefore, angle BCG = angle DFE.
But angle DFE = angle BCA;

therefore, angle BCG = BCA,
the less angle to the greater, which is impossible.
Therefore the line AB is not unequal to the line DE; therefore it is equal.
And BC = EF.
Therefore, there are two sides, AB and BC respectively equal to two sides DE and EF, and angle ABC = angle DEF.
Therefore, the base AC = base DF and the remaining angle BAC = the remaining angle EDF.
Again, let it have been given that the sides subtending the equal angles are equal to one another; let the side AB = DE.

Then again I say that the other sides of the one shall be respectively equal to the corresponding sides of the other; that is, AC = DF and BC = EF; and the remaining angle of the one shall be equal to the remaining angle of the other; that is, angle CAB = angle EDF.

For if the side BC is not equal to side EF, then one of them is greater. Let the side BC, if it be possible, have been greater.
Let there have been cut off equal to EF the line BH, and let there have been drawn the line HA.
Now since BH = EF, and AB = DE; therefore the two sides BH and AB are respectively equal to the two sides EF and DE, and they contain equal angles.
Therefore, base AH = base DF, and triangle BHA = triangle EFD, and the remaining angles are respectively equal to the remaining angles.
Therefore, angle BHA = angle EFD.
But angle EFD = angle BCA.
Therefore, angle BHA = angle BCA.
Therefore, an exterior angle of triangle AHC, namely angle BHA, is equal to an interior opposite angle, namely the angle BCA,
which is impossible.
Therefore, the side BC is not unequal to the side EF, therefore it is equal to it.
And the side AB = side DE.
Therefore, the two sides, AB and BC are respectively equal to the two sides DE and EF, and they contain equal angles.
Therefore, base AC = base DF, and triangle ABC = triangle DEF, and the remaining angle BAC = the remaining angle EDF.

If therefore two triangles have two angles of the one respectively equal to two angles of the other, and have also one side of the one equal to one side of the other, either the side that lies between the two equal angles, or the side that is subtended by one of the equal angles, the other sides of the one shall be respectively equal to the other sides of the other, and the other angle of the one shall be equal to the other angle of the other.

The very thing it was required to demonstrate.

If a straight line falling on two straight lines makes the alternate angles equal to one another, those two straight lines are parallel to one another.

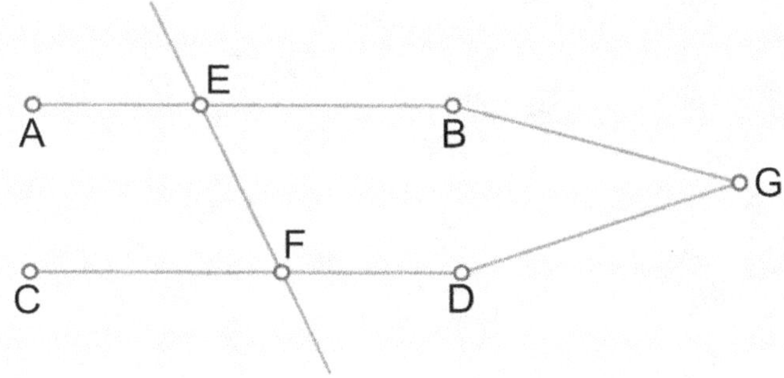

Let there have been given, falling upon two straight lines, namely AB and CD, a straight line, EF; making the alternate angles equal to one another, namely angle AEF = angle EFD.
I say that AB is parallel to CD.
For if not, the lines extended continually shall meet together, either on the side of B and D or on the side of A and C. Let them have been extended, and let them meet, if it be possible, on the side of B and D, at point G.
Therefore, in the triangle GEF, the exterior angle AEF = the opposite interior angle EFG, which is impossible.
Therefore the sides AB and CD being extended on the side of B and D shall not meet.
In like manner also may it be proved that they shall not meet on the side of A and C.
But lines that being extended on both sides do not meet are parallel lines.
Therefore, AB is parallel to CD.

If therefore a straight line falling on two straight lines makes the alternate angles equal to one another, those two straight lines are parallel to one another.

The very thing it was required to demonstrate.

Proposition 28

If a straight line falling on two straight lines make the exterior angle equal to the
opposite interior angle on the same side, or the interior angles on the same side
equal to two right angles, those two straight lines shall be parallel to one another.

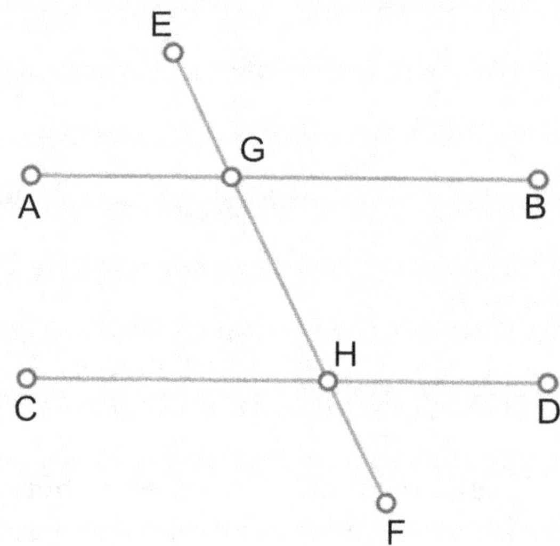

Let there have been given, falling upon two straight lines, namely AB and CD, a straight line, EF;
making the exterior angle EGB equal to the interior opposite angle on the same side GHD, or
making the interior angles on the same side, BGH and GDH, together equal to two right angles.
I say the line AB is parallel to the line CD.
For given angle EGB = GHD,
and angle EGB = angle AGH;
therefore, angle AGH = angle GHD, and they are alternate angles.
Therefore AB is parallel to CD.
Again given angles BGH and GHD together are equal to two right angles,
and angles AGH and BGH together are equal to two right angles,
therefore angles AGH and BGH together are equal to BGH and GHD together.
Take away the angle BGH that is common to both,
and the remaining angle AGH = remaining angle GHD, and they are alternate angles.
Therefore, AB is parallel to CD.

If therefore a straight line falling on two straight lines make the exterior angle
equal to the opposite interior angle on the same side, or the interior angles on the
same side equal to two right angles, those two straight lines shall be parallel to one
another.

The very thing it was required to demonstrate.

A straight line falling on two parallel lines makes the alternate angles equal to one another, and makes the exterior angle equal to the opposite interior angle on the same side, and makes the interior angles on the same side equal to two right angles.

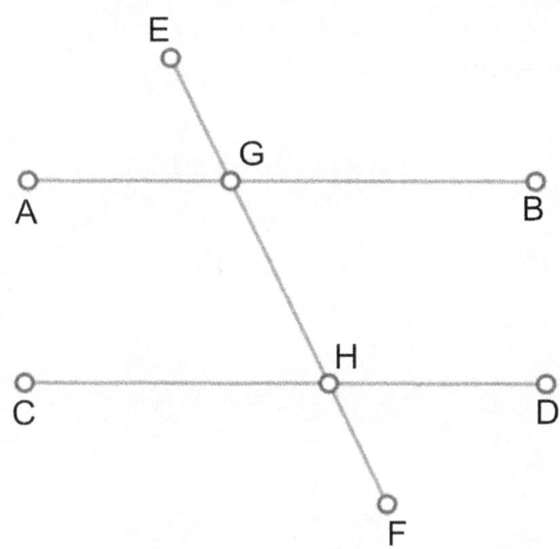

Let there have been given parallel lines AB and CD and falling upon them the line EF.
I say that the alternate angles, AGH and GHD, are equal to one another and the exterior angle EGB = the interior opposite angle on the same side GHD, and the interior angles on the same side, BGH and GHD, together are equal to two right angles.
For if the angle AGH be not equal to GHD, then one of them is greater. Let it have been that AGH is the greater.
And since angle AGH is greater than angle GHD, add the angle BGH to both.
 Therefore the angles AGH and BGH together are greater than the angles GHD and BGH together.
But angles AGH and BGH together are equal to two right angles.
Therefore, angles GHD and BGH together are less than two right angles.
But when a straight line falling upon two straight lines makes on one and the same side, the two interior angles less than two right angles, then shall these two straight lines being extended at length meet on that side in which are the two angles less than two right angles.
Therefore the lines AB and CD being extended indefinitely will at length meet.
But they cannot meet because they are parallel.
Therefore the angle AGH is not unequal to the angle GHD,
therefore it is equal.
And angle AGH = angle EGB.
Therefore, angle EGB = angle GHD.
Add the angle BGH to both;

therefore, angles EGB and BGH together are equal to angles BGH and GHD together.
But the angles EGB and BGH together are equal to two right angles.
Therefore, the angles BGH and GHD together are also equal to two right angles.

Therefore a straight line falling on two parallel lines makes the alternate angles equal to one another, and makes the exterior angle equal to the opposite interior angle on the same side, and makes the interior angles on the same side equal to two right angles.

The very thing it was required to demonstrate.

Proposition 30

Straight lines parallel to the same straight line are parallel to one another.

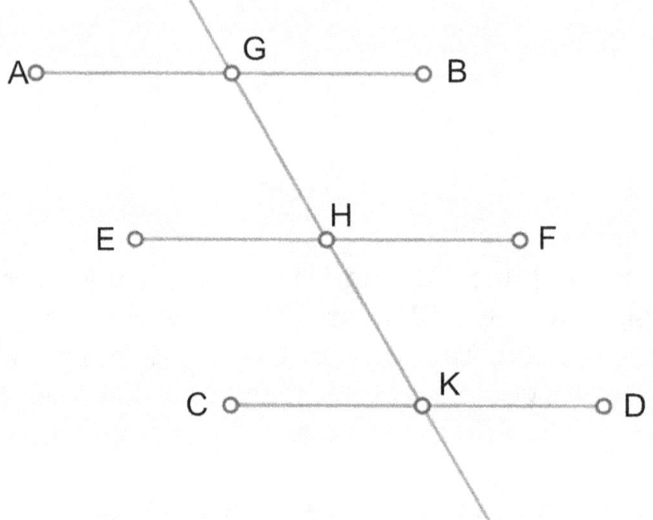

Let there have been given the straight lines AB and CD parallel to the straight line EF.
I say that the line AB is parallel to the line CD.
Let there have fallen upon the three lines a straight line GHK.
For since the line GHK falls upon the parallel lines AB and EF,
therefore the angle AGH = angle GHF.
Again, since the line GK falls upon the parallel lines EF and CD,
therefore the angle GHF = angle GKD.
Now it is demonstrated that angle AGH = angle GHF and angle GKD = GHF.
Therefore angle AGK = angle GKD, and they are alternate angles.
Therefore AB is parallel to CD.

Therefore straight lines parallel to the same straight line are parallel to one another.

The very thing it was required to demonstrate.

Through a given point to draw a straight line parallel to a given straight line.

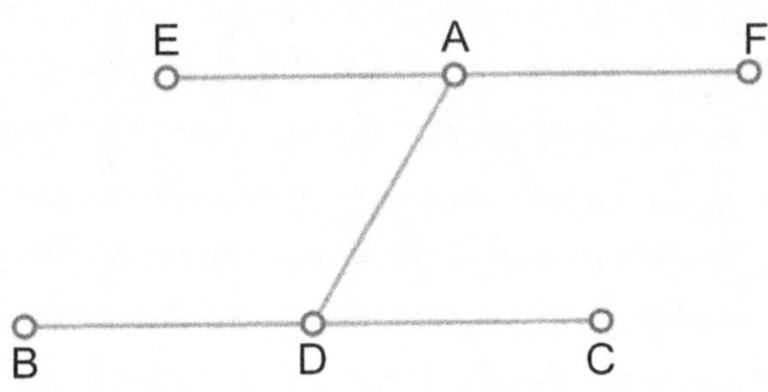

Let there have been given a point A, and a straight line BC.
It is require to draw through the point A, a straight line parallel to BC.
Let a point have been taken at random on line BC, point D.
Let there have been drawn the line AD.
Upon the given line AD, at the given point A, let there have been constructed an angle equal to the given angle ADC, namely angle DAE.
And taking line EA let it have been extended in a straight line by AF.
For since the straight line AD falls upon the straight lines BC and EF making the alternate angles equal, namely angle EAD = angle ADC;
therefore, EF is parallel to BC.

Therefore through a given point, A, is drawn a straight line, line EAF, parallel to a given straight line, BC.

The very thing it was required to demonstrate.

Proposition 32

If one of the sides of a triangle be extended, the exterior angle produced is equal
to the two interior and opposite angles together. And the three angles of a triangle
together are equal to two right angles.

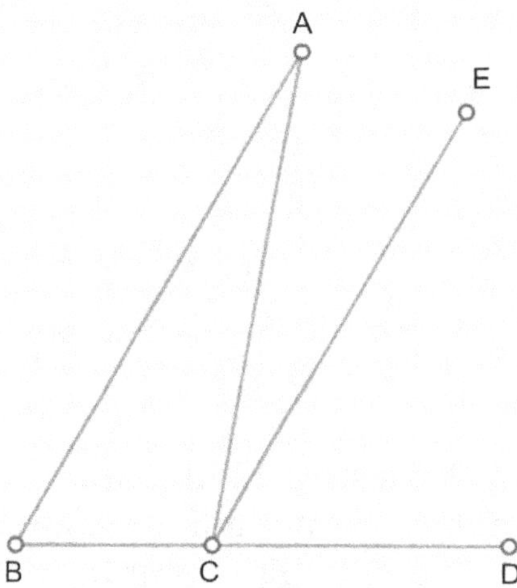

Let there have been given a triangle, triangle ABC, with one side extended, namely side BC
extended to D.
I say that the exterior angle, ACD, is equal to the two opposite interior angles, CAB and ABC
together; and the three interior angles of the triangle, CAB, ABC, and BCA, together are equal to
two right angles.
For, through the given point C, draw a straight line parallel to the given straight line AB, the line
CE.
And since AB is parallel to CE, and upon them falls the straight line AC;
therefore the alternate angles, BAC and ACE are equal.
Again, since AB is parallel to CE, and upon them falls the straight line BD, the exterior angle
ECD = the opposite interior angle on the same side ABC.
And it is demonstrated that angle ACE = angle BAC;
Therefore, the whole outward angle ACD = the two opposite interior angles BAC and ABC
together.
Add the angle ACB to them both;
therefore, the angles ACD and ACB together are equal to the three angles ABC, BCA, and BAC
together.
But the angles ACD and ACB are equal to two right angles;
therefore, the angles ABC, BCA, and BAC together are equal to two right angles.

If therefore one of the sides of a triangle be extended, the exterior angle produced
is equal to the two interior and opposite angles together. And the three angles of a
triangle together are equal to two right angles.

The very thing it was required to demonstrate.

Proposition 33

Two straight lines joining together, on the same side, two equal parallel lines are also equal and parallel.

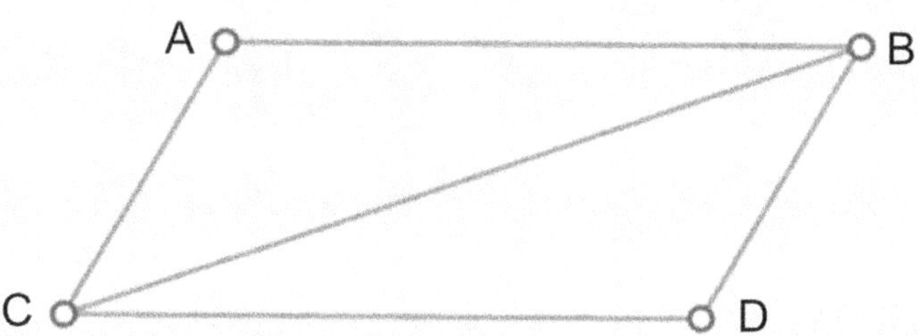

Let there have been given two equal and parallel straight lines, AB and CD. And let them have been drawn on the same side by lines AC and BD.
I say that AC and BD are also equal and parallel to one another.
Let the line BC have been drawn.
And since AB is parallel to CD, and upon them falls the straight line BC; therefore the alternate angles, ABC and BCD, are equal to one another.
And since AB = CD, and BC is common, the two lines AB and BC are respectively equal to the two lines CD and BC, and the angle ABC = angle BCD;
therefore, the base BD = the base AC, and triangle ABC = triangle DCB, and the remaining angles are equal to the remaining angles, namely those that are subtended by equal sides.
Therefore, angle ACB = angle CBD, and angle BAC = angle BDC.
And since on the two straight lines AC and BD there falls a straight line BC making the alternate angles, ACB and CBD, equal to one another, the line AC is parallel to BD.
And it was demonstrated that they were equal.

Therefore two straight lines joining together, on the same side, two equal parallel lines are also equal and parallel

The very thing it was required to demonstrate.

In parallelograms, the sides and angles that are opposite to one another are equal to one another; and the diameter divides the parallelogram into two equal parts.

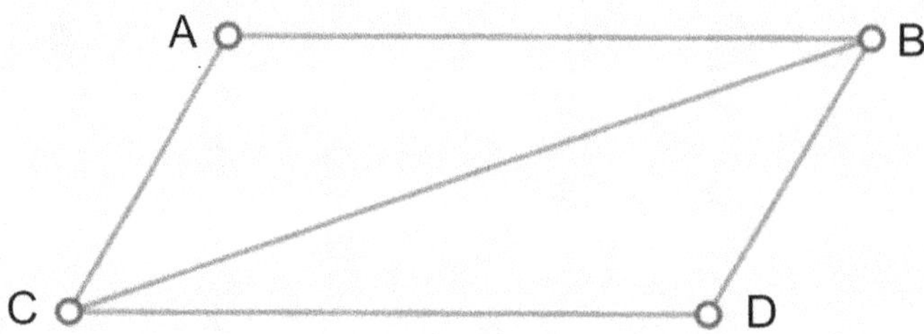

Let there have been given a parallelogram, ABCD; and let the diameter have been BC.
I say that opposite sides and angles of the parallelogram ABCD are equal and that the diameter BC divides it into two equal parts.
For since AB is parallel to CD, and upon them falls the straight line BC, the alternate angles ABC and BCD are equal to one another.
Again, since AC is parallel to BD, and upon them falls the straight line BC, the alternate angles ACB and CBD are equal to one another.
Now, there are two triangles, ABC and BCD having two angles of one, namely ABC and ACB, respectively equal to two angles of the other, BCD and CBD, and one side equal to one side, the side that lies between the two equal angles, which side is common to both, side BC;
therefore, the other sides remaining are respectively equal to the other sides, and the angle remaining is equal to the angle remaining.
Therefore, side AB = side CD; side AC = side BD; and angle BAC = angle BDC.
And since angle ABC=BCD, and angle CBD = ACB;
therefore, the whole angle ABD = the whole angle ACD.
And it is demonstrated that angle BAC = angle BDC.

Therefore in parallelograms, the sides and angles that are opposite to one another are equal to one another.

I say also that the diameter divides it into two equal parts.
For since AB = CD, and BC is common, therefore the two AB and BC are respectively equal to the two CD and BC, and the angle ABC = angle BCD.
Therefore the base AC = the base BD, and the triangle ABC = triangle BCD.

Therefore the diameter BC divides the parallelogram ABCD into two equal parts.

The very thing it was required to demonstrate.

Proposition 35

Parallelograms standing upon the same base and in the same parallels are equal to one another.

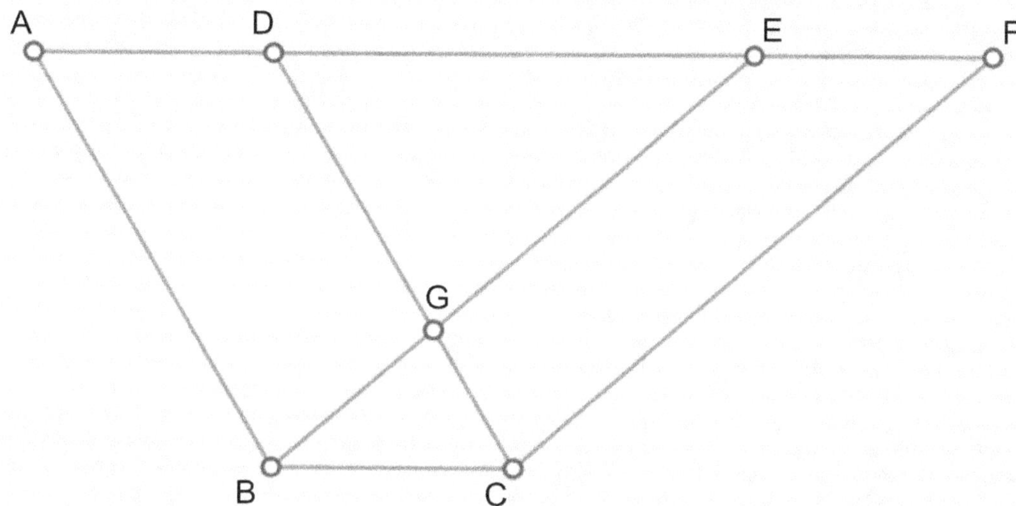

Let there have been given two parallelograms, ABCD and EBCF, that stand upon the same base, line BC, and in the same parallels, AF and BC.
I say that parallelogram ABCD = parallelogram EBCF.
For since ABCD is a parallelogram, side AD = side BC.
For the same reason, side EF = side BC.
Therefore, AD = EF.
Add DE to both; therefore, the whole AE = the whole DF.
And the side AB = the side DC.
Therefore, the two EA and AB are respectively equal to the two DF and DC, and angle FDC = angle EAB, namely the exterior angle to the opposite interior angle.
Therefore, the base EB = the base CF, and the triangle EAB = the triangle FDC.
Take away the triangle DGE which is common to both; therefore, the remainder, trapezium ABGD = the remainder, trapezium EGCF.
Add the triangle GBC to both; therefore the whole ABCD = the whole EBCF.

Therefore parallelograms standing upon the same base and in the same parallels are equal to one another.

The very thing it was required to demonstrate.

Parallelograms standing on equal bases and in the same parallels are equal to one another.

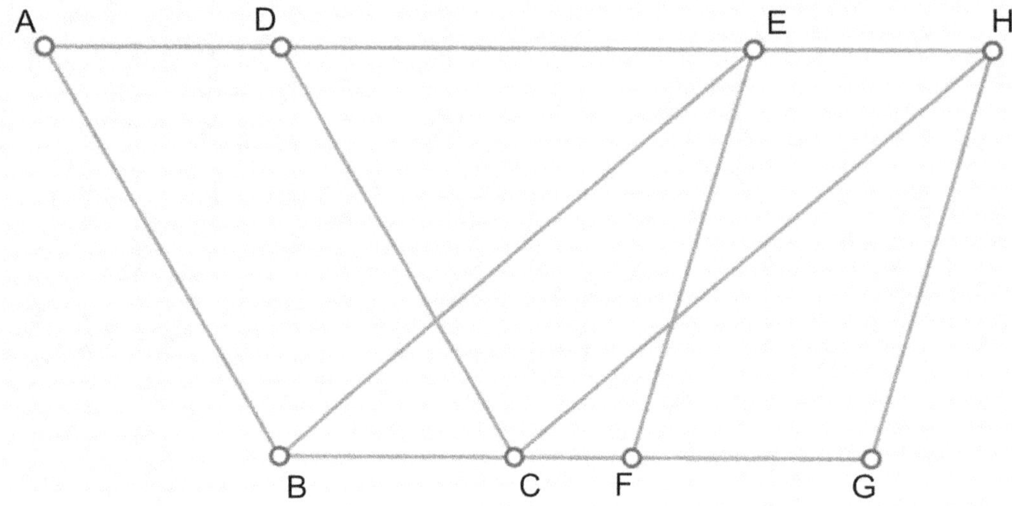

Let there have been given the parallelograms ABCD and EFGH standing upon equal bases, the bases BC and FG, and in the same parallels, lines AH and BG.
I say that the parallelogram ABCD equals parallelogram EFGH.
Let there have been drawn a straight line from B to E, and another from C to H.
Since BC = FG, but FG = EH;
therefore BC = EH
and they are parallel lines;
and the lines BE and CH join them together.
But two straight lines joining two equal parallel lines are also equal and parallel.
Therefore EBCH is a parallelogram,
and it is equal to parallelogram ABCD, for they both have the same base, BC, and are in the same parallels, BC and EH.
And for the same reason, the parallelogram EFGH = the parallelogram EBCH.
Therefore, parallelogram ABCD = parallelogram EFGH.

Therefore parallelograms standing on equal bases and in the same parallels are equal to one another.

The very thing it was required to demonstrate.

Triangles standing upon the same base and in the same parallels are equal to one another.

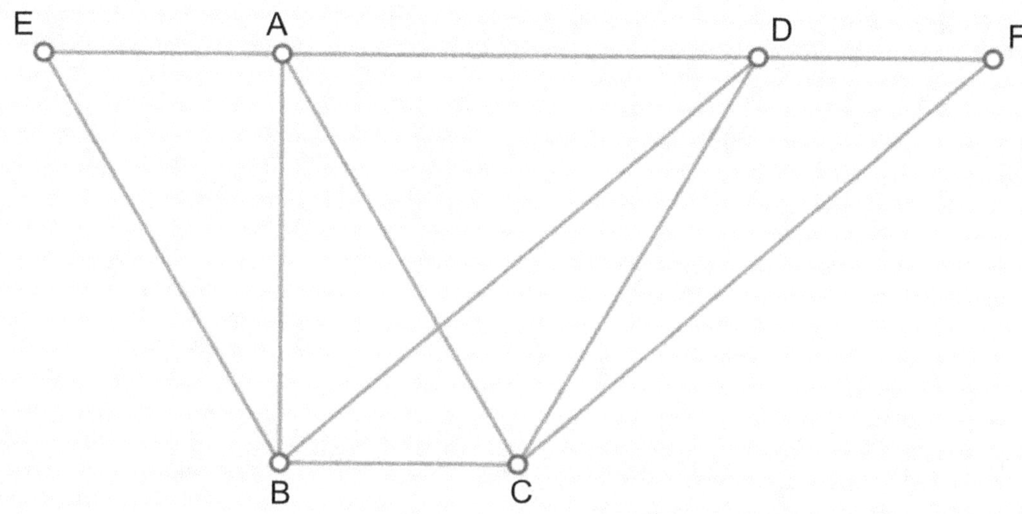

Let there have been given triangles ABC and DBC, standing upon the same base, BC, and in the same parallel lines, AD and BC.
I say that triangle ABC = triangle DBC.
Let the line AD have been extended continually in both directions to points E and F such that through point B, let there have been drawn a line parallel to AC, line EB; and through point C let there have been drawn a line parallel to BD, line CF.
Therefore, EBCA and DBCF are parallelograms,
and they are equal because they stand upon the same base BC, and are in the same parallels, BC and EF.
But the triangle ABC is half of the parallelogram EBCA because the diameter AB cuts it into two equal parts.
But the triangle DBC is half of the parallelogram DBCF because the diameter DC cuts it into two equal parts.
Therefore, triangle ABC = triangle DBC.

Therefore triangles standing upon the same base and in the same parallels are equal to one another.

The very thing it was required to demonstrate.

Triangles standing upon equal bases in the same parallels are equal to one another.

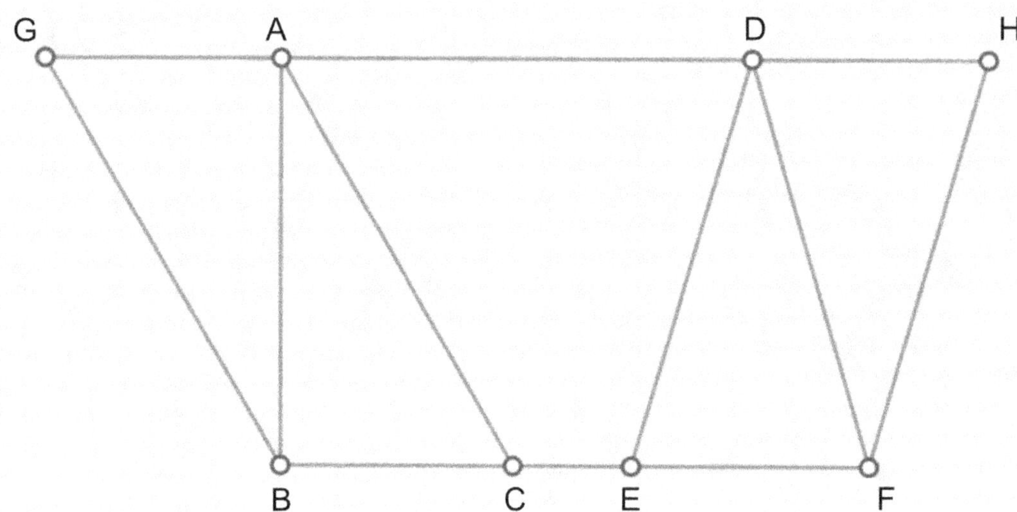

Let there have been given the triangles ABC and DEF, standing upon equal bases, BC and EF, and in the same parallels, BF and AD.
I say that triangle ABC = triangle DEF.
Let the line AD have been extended continually in both directions to points G and H such that through point B, let there have been drawn a line parallel to AC, line GB; and through point F let there have been drawn a line parallel to ED, line FH.
Therefore, GBCA and DEFH are parallelograms.
But the parallelogram GBCA = the parallelogram DEFH for they stand upon equal bases, BC and EF, and are in the same parallels, BF and GH.
But triangle ABC is half of parallelogram GBCA, for the diameter AB cuts it into two equal parts. And the triangle DEF is half of parallelogram DEFH, for the diameter FD cuts it into two equal parts.
Therefore, triangle ABC = triangle DEF.

Therefore triangles standing upon equal bases in the same parallels are equal to one another.

The very thing it was required to demonstrate.

Equal triangles standing upon the same base, and on the same side, are in the same parallels.

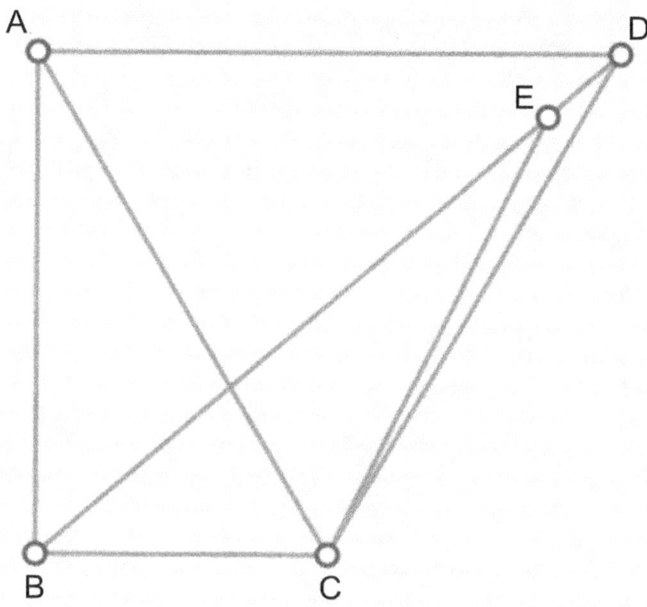

Let there have been given two equal triangles, ABC and DBC, standing upon the same base BC and on the same side.
I say that they are in the same parallels.
Let there have been drawn a straight line joining point A and point D.
I say that AD is parallel to BC.
For if not, then through point A let there have been drawn a line parallel to line BC, the line AE (E being the point where the new parallel line intersects line AD).
Let there have been drawn a straight line joining C and E.
Therefore the triangle EBC = the triangle ABC, for they stand upon the same base, BC, and are in the same parallels, AE and BC.
But the triangle DBC = triangle ABC.
Therefore, the triangle DBC = the triangle EBC, the greater equal to the less,
which is impossible.
Therefore the line AE is not parallel to line BC.
And in like manner it may be proved that no other line besides AD is parallel to BC.
Therefore, AD is parallel to BC.

Therefore equal triangles standing upon the same base, and on the same side, are in the same parallels.

The very thing it was required to demonstrate.

Proposition 40

Equal triangles standing upon equal bases, and on the same side, are in the same parallels.

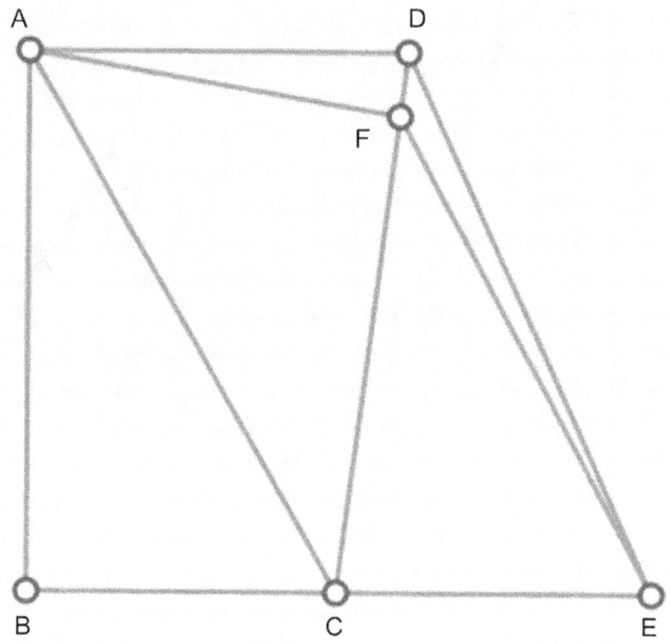

Let there have been given equal triangles, ABC and CDE standing upon equal bases, BC and CE, and on the same side.
I say that they are in the same parallels.
Let there have been drawn a line from point A to point D.
I say that AD is parallel to BE.
For if not, then let there have been drawn through point A a line parallel to BE, the line AF (F being the point where the new parallel line intersects line CD).
And let there have been drawn a line from point F to point E.
Therefore, the triangle BAC = the triangle CFE, for they stand upon equal bases in the same parallels, BE and AF.
But the triangle ABC = triangle CDE.
Therefore, triangle CDE = triangle FCE, the greater equal to the less,
which is impossible.
Therefore, AF is not parallel to BE.
And in like manner it may be demonstrated that no other line besides AD is a parallel line to BE.
Therefore AD is parallel to BE.

Therefore, equal triangles standing upon equal bases, and on the same side, are in the same parallels.

The very thing it was required to demonstrate.

If a parallelogram and a triangle have the same base and are in the same parallels, the parallelogram is double of the triangle.

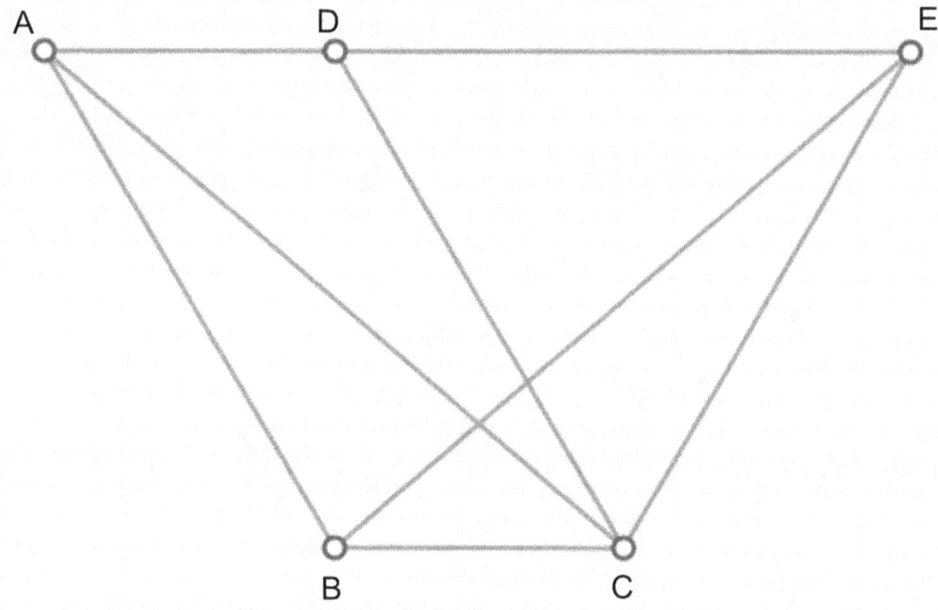

Let there have been given a parallelogram ABCD and the triangle EBC having the same base, BC, and let them have been in the same parallels, BC and AE.
I say that the parallelogram ABCD is double the triangle EBC.
Let there have been drawn a straight line from point A to point C.
Therefore, triangle ABC = triangle EBC because they stand upon the same base, BC, and are in the same parallels, AE and BC.
But the parallelogram ABCD is double the triangle ABC for the diameter of it, AC, cuts it into two equal parts.
Therefore, the parallelogram ABCD is double of the triangle EBC.

Therefore, if a parallelogram and a triangle have the same base and are in the same parallels, the parallelogram is double of the triangle.

The very thing it was required to demonstrate.

To construct a parallelogram equal to a given triangle in an angle equal to a given rectilineal angle.

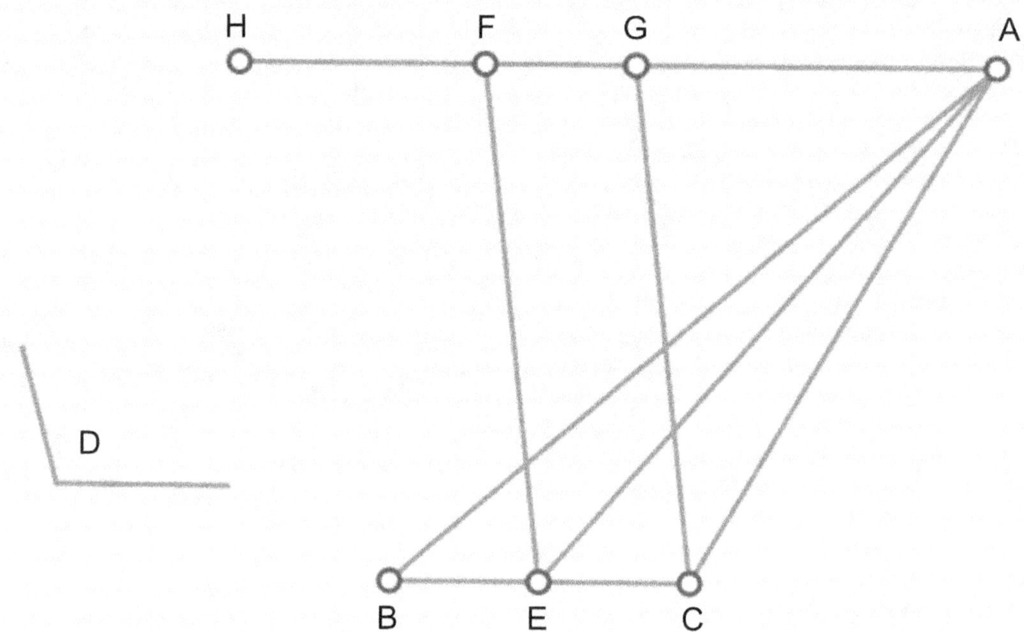

Let there have been given a triangle ABC, and let there have been given a rectilineal angle D. It is required that equal to ABC there be constructed a parallelogram in an angle equal to D.
Let the line BC have been bisected at E.
And let the line AE have been joined.
And upon the line EC, let there have been constructed at the point E, an angle equal to angle D, the angle CEF.
And through the point A let there have been drawn a line parallel to the line BC, the line AH (the point F being the intersection of AH and EF).
And through the point C, let there have been drawn a line parallel to the line EF, the line CG (G being the intersection of AH and CG).
Therefore, FECG is a parallelogram.
And since BE = EC, the triangle ABE = the triangle AEC, for they stand upon equal bases in the same parallels, BC and AH.
Therefore, triangle ABC is double triangle AEC.
And parallelogram CEFG is also double the triangle AEC, for they have the same base, EC, and are in the same parallels BC and AH.
Therefore parallelogram CEFG = triangle ABC, and has the angle CEF = given angle D.

Therefore, a parallelogram is constructed equal to a given triangle, ABC, in an angle, CEF, equal to a given rectilineal angle, D.

The very thing it was required to do.

In every parallelogram, the complements of the parallelograms about the diameter are equal to one another.

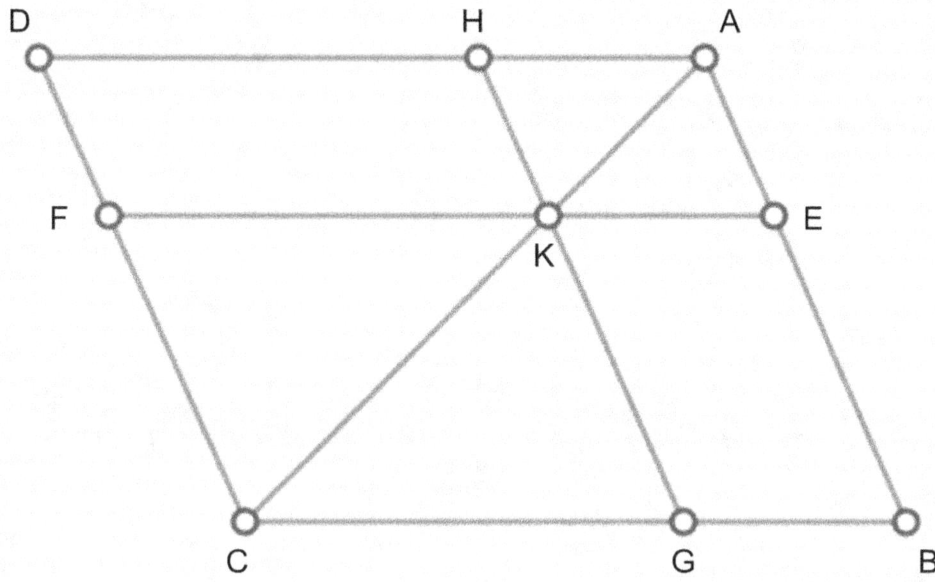

Let there have been given a parallelogram, ABCD, with the diameter AC; and about the diameter let there have been constructed the parallelograms EH and GF; and let the complements be BK and KD.
I say that the complement BK = the complement KD.
Since ABCD is a parallelogram and the diameter is AC,
triangle ABC = triangle ADC.
Again, since AEKH is a parallelogram and the diameter is AK,
triangle AEK = triangle AHK.
And for the same reason,
triangle KGC = triangle KFC.
Now since AEK = AHK and KGC = KFC, the triangles AEK and KCG together are equal to triangles AHK and KFC together; and the whole triangle ABC = the whole triangle ADC.
Therefore, the remainder, the complement BK = the remainder, the complement KD.

Therefore, in every parallelogram, the complements of the parallelograms about the diameter are equal to one another.

The very thing it was required to demonstrate.

Proposition 44

Upon a given straight line, to apply a parallelogram equal to a given triangle, and containing an angle equal to a given rectilineal angle.

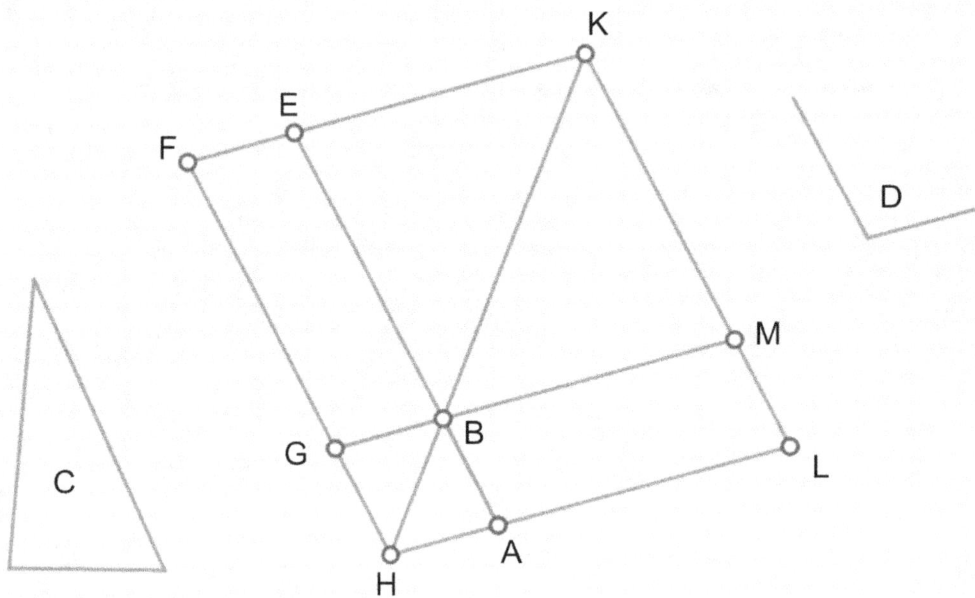

Let there have been given a straight line AB, and let there have been given a triangle C, and a rectilineal angle D.
It is required upon the given line AB, to apply a parallelogram equal to the given triangle C and containing an angle equal to the given angle D.
Let there have been constructed a parallelogram BGEF equal to the triangle C in an angle, BGF, equal to the angle D.
And to the line EB, let there have been joined the line AB, in such a way that they both make one straight line.
And let the line FG have been extended to H.
And through point A, let there have been drawn a line parallel to either BG or EF, line AH.
And let the point H have been joined to the point B.
And since upon the parallel lines AH and EF, there falls a certain straight line HF, therefore the angles AHF and HFE together are equal to two right angles.
Therefore, the angles BHG and GFE together are less than two right angles.
But, when a straight line falling upon two straight lines makes on one and the same side, the two interior angles less than two right angles, then shall these two straight lines being extended indefinitely meet on that side in which are the two angles less than two right angles.
Therefore, the two straight lines HB and FE being indefinitely extended will at length meet.
Let them have been extended, and let them have met at a point, the point K.
And through point K, let there have been drawn a line parallel to either EA or FH, the line KL.
And let the lines HA and GB have been extended until they intersect the line KL at points L and M (forming line HAL and line GBM).

Therefore HKLF is a parallelogram, and the diameter of it is HK;
and about the diameter HK are the parallelograms AG and ME, and remaining are the
complements LB and BF;
therefore, the complement LB = the complement BF.
But the parallelogram BF is equal to the triangle C.
And since line FH is parallel to line KL, and upon them lies line GM;
therefore, angle FGB = angle BML.
But angle FGB = angle D;
therefore, angle BML = angle D.

Therefore, upon a given straight line, AB, is applied a parallelogram, LB, equal to a
given triangle, C, and containing an angle, BML, equal to a given rectilineal angle, D.

The very thing it was required to do.

Proposition 45

To describe a parallelogram equal to any rectilineal figure and containing an angle
equal to a given rectilineal angle.

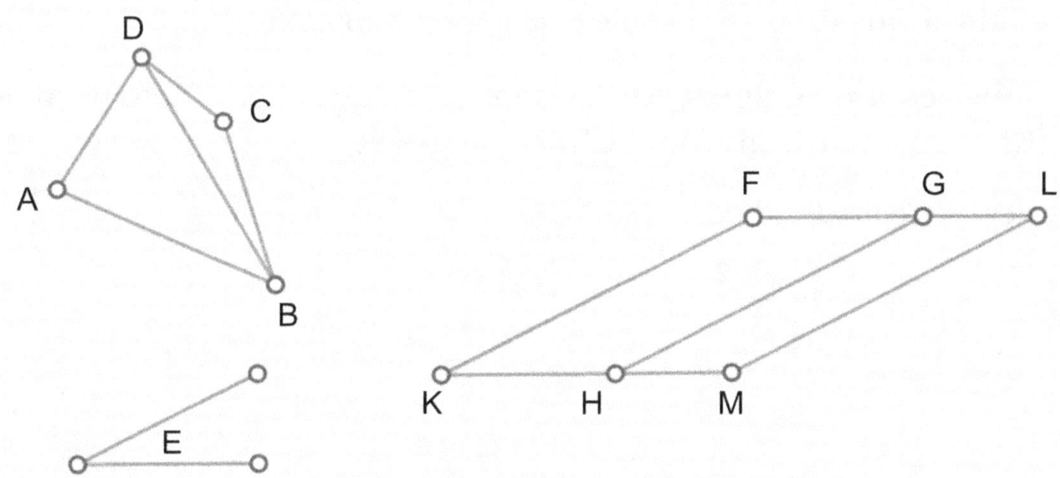

Let there have been given a rectilineal figure ABCD, and let there have been given a rectilineal
angle E.
It is required to describe a parallelogram equal to the rectilineal figure ABCD and containing an
angle equal to angle E.
Let a straight line have been drawn from B to D.
Let there have been constructed a parallelogram equal to triangle ABD, the parallelogram FH,
having angle FKH equal to angle E.
And upon the line GH, let a parallelogram, GM, be applied equal to triangle DBC, and having the
angle GHM equal to angle E.

And since both of the angles HKF and GHM are equal to angle E;
angle HKF = angle GHM.
Add the angle KHG to both;
therefore, FKH and KHG together are equal to KHG and GHM together.
But the angles FKH and KHG together are equal to two right angles;
therefore, angles KHG and GHM together are equal to two right angles.
Now, unto a straight line, GH, and at a point on it, H, are drawn two straight lines, KH and HM,
not both on the same side, making adjacent angles equal to two right angles.
Therefore, the lines KH and HM are in one straight line.
And since upon the parallel lines KM and FG falls the straight line HG, the alternate angles,
MHG and HGF are equal to one another.
Add the angle HGL to both; therefore, the angles MHG and HGL together are equal to the angles
HGF and HGL together.
But the angles MHG and HGL are equal to two right angles.
Therefore the lines FG and GL make one straight line.
And since KF = HG, and it is also parallel to it;
and HG = ML, and it is also parallel to it;
therefore, KF = ML, and it is also parallel to it.
But the straight lines KM and FL join them together;
therefore, KM = FL, and they are parallel lines.
Therefore, KMFL is a parallelogram.
And since triangle ABD = parallelogram FH, and triangle DCB = parallelogram GM; therefore,
the whole rectilineal figure ABCD = the whole parallelogram KFLM.

Therefore, to the rectilineal figure ABCD is made an equal parallelogram KFLM,
with angle FKM equal to the given rectilineal angle E.

The very thing it was required to do.

Upon a given straight line, to describe a square.

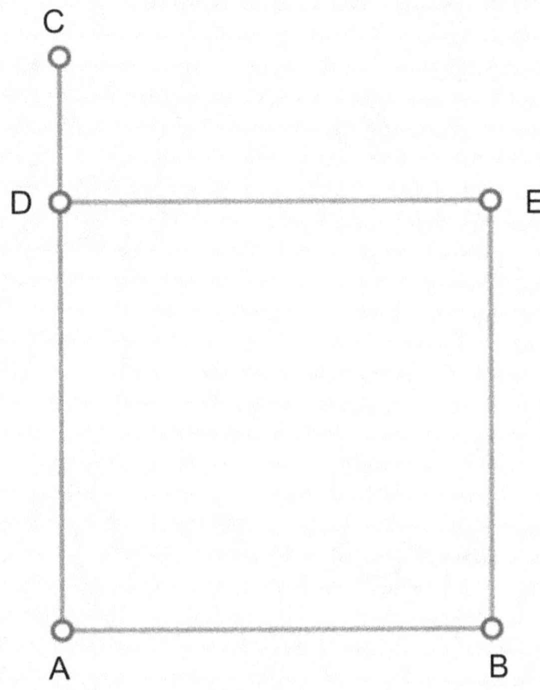

Let there have been given a straight line, AB.
It is required upon the line AB to describe a square.
Upon the line AB, and from a point on it, A, let there have been raised up a perpendicular line AC,
And (on AC) let there have been cut off a line equal to AB, line AD.
And from point D, let there have been drawn a line parallel to AB, line DE.
And from point B, let there have been drawn a line parallel to AD, line BE.
Therefore, ADEB is a parallelogram;
therefore, AB = DE, and AD = BE.
But AB = AD;
therefore, the four lines AB, AD, DE, EB are equal to one another.
Therefore, the parallelogram ADEB consists of equal lines.
I say also that it is right angled.
For since upon the parallel lines, AB and DE, falls a straight line AD,
angle BAD and ADE together equal two right angles.
But angle BAD is a right angle; therefore, ADE is a right angle.
But in parallelograms opposite sides and angles are equal to one another.
Therefore, the two opposite angles, ABE and BED are each of them a right angle.
Therefore the parallelogram ABED is right angled and it is also demonstrated equilateral.

Therefore it is a square, and it is described upon the given line AB.

The very thing it was required to do.

In right triangles, the square of the side that subtends the right angle is equal to the squares of the sides containing the right angle.

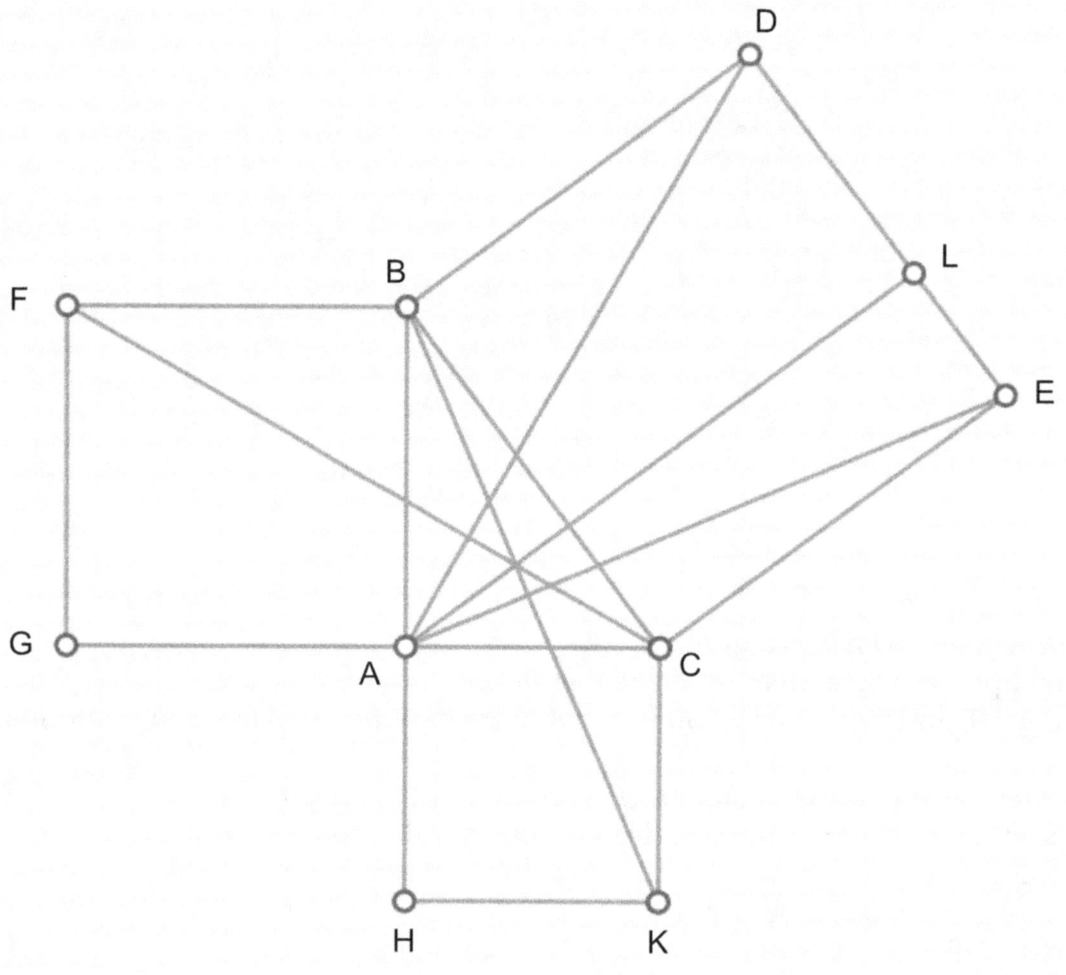

Let there have been given a right triangle, ABC, having BAC a right angle.
I say the square of the line BC = the squares of the lines AB and AC.
Upon the line BC, let there have been described a square, square BDEC.
And upon the lines BA and AC let there have been described the squares BFGA and AHKC.
And through point A, let there have been drawn a line parallel to either BD or CE, the line AL (L being the intersection of AL and DE).
And let the lines AD and CF have been drawn.
And since the angles BAC and BAG are right angles, therefore, unto a straight line, BA, and at a point on it A, are drawn two straight lines, AC and AG, not both on the same side, making the two angles together equal to two right angles.
Therefore, the two lines AC and AG make one straight line.
And for the same reason, BA and AH make one straight line.
And since angle DBC = angle FBA, add the angle ABC to both;

therefore, the whole angle DBA = whole angle FBC.
And since the two lines AB and BD are respectively equal to the two lines BF and BC, and angle DBA = angle FBC; therefore, the base AD = the base FC and triangle ABD = triangle FBC.
But the parallelogram BL is double of triangle ABD, for they both have the same base, BD, and are in the same parallels, BD and AL.
And the square GB is double the triangle FBC, for they both have the same base, BF, and are in the same parallels, FB and GC.
But the doubles of equal things are equal; therefore, the parallelogram BL = the square GB.
And in like manner, let the lines AE and BK have been drawn, and it may be demonstrated that the parallelogram CL = the square HC.
Therefore, the whole figure BDEC = the two squares GB and HC together.
But the square BDEC is described upon the line BC, and the squares GB and HC are described upon the lines BA and AC;
therefore, the square upon BC = the squares upon BA and AC.

Therefore, in right triangles, the square of the side that subtends the right angle is equal to the squares of the sides containing the right angle.

The very thing it was required to demonstrate.

If the square on one of the sides of a triangle be equal to the squares on the other two sides of the same triangle, the angle contained by those other two sides is a right angle.

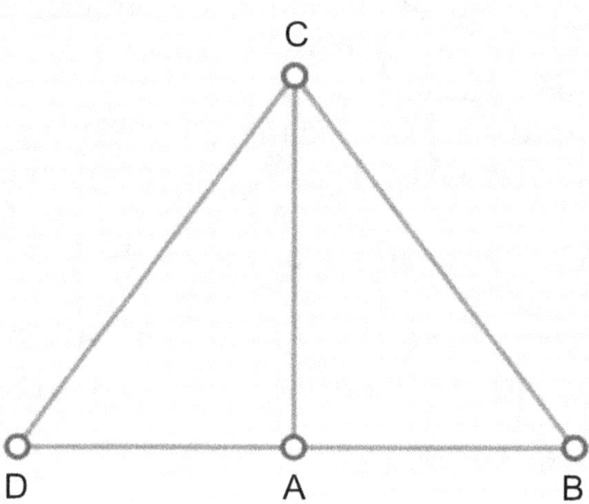

Let there have been given a triangle, ABC, and let the square of one side, BC, have been equal to the squares on BA and BC.
I say that the angle BAC is a right angle.
Let there have been raised up from point A, a line perpendicular to AC, the line AD.
And let the line AD have been cut off equal to AB.
And let the line DC have been drawn.
And since DA = AB, the square on DA = the square on AB.
Add the square of line AC to both.
Therefore, the squares on DA and AC = the squares on BA and AC.
But the square on DC = the squares on AD and AC, and the square on BC = the squares on BA and AC.
Therefore, the square on DC = the square on BC; therefore, the side DC = the side BC.
And since AB = AD, and AC is common, the two sides DA and AC are respectively equal to the two sides BA and AC, and the base DC = the base BC; therefore, the angle DAC = the angle BAC. But the angle DAC is a right angle; therefore the angle BAC is a right angle.

Therefore, if the square on one of the sides of a triangle be equal to the squares on the other two sides of the same triangle, the angle contained by those other two sides is a right angle.

The very thing it was required to demonstrate.

Book II

Definitions

1. Every rectangle is said to be contained by two straight lines containing a right angle.
2. In every parallelogram, one of the parallelograms about the diameter, whichever one it be, together with the two complements is called a gnomon.

If there be two straight lines, and if one of them be divided into any number of parts, the rectangle contained by the two straight lines is equal to the rectangles that are contained by the undivided line and every one of the parts of the other line.

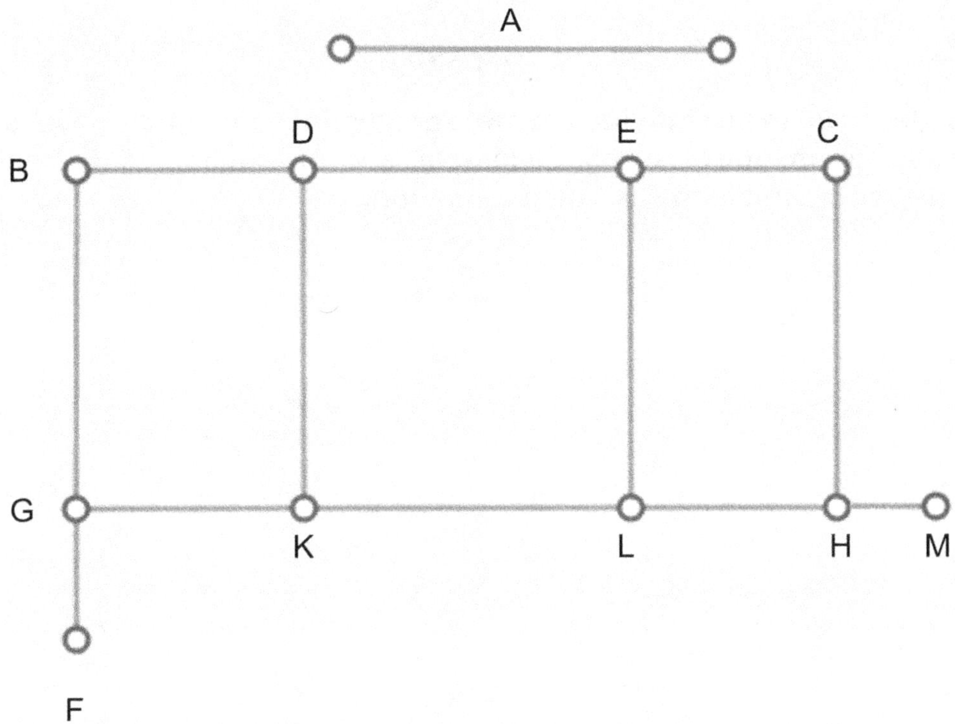

Let there have been two straight lines, A, and BC,
and let one of them, namely BC, have been divided randomly at the points D and E.
I say that the rectangle contained by A and BC = the rectangle contained by A and BD, the rectangle contained by A and DE, and the rectangle contained by A and EC.
For from the point B on the line BC, raise up a line perpendicular to BC, the line BF.
And let a line equal to A have been cut off from BF, the line BG.
And through the point G, let there have been drawn a line parallel to line BC, the line GM.
And through the points D, E, and C, let there have been drawn lines parallel to BG, namely DK, EL, and CH.
Now, the rectangle BH is equal to the sum of the rectangles BK, DL, and EH.
But the rectangle BH is equal to that contained by the lines A and BC,
since it is contained by BC and BG and BG = A.

And the rectangle BK is equal to that contained by A and BD, since it is contained by BG and BD and BG = A.

And the rectangle DL is equal to that contained by A and DE, since it is contained by DK and DE, and DE = BG which is equal to A.

And similarly, the rectangle EH is equal to that contained by A and EC, since it is contained by EL and EC, and EC = BG which is equal to A.

Therefore, that which is contained by A and BC = that contained by A and BD, that contained by A and DE, and that contained by A and EC.

If therefore, there be two straight lines, and if one of them be divided into any number of parts, the rectangle contained by the two straight lines is equal to the rectangles that are contained by the undivided line and every one of the parts of the other line.

The very thing it was required to demonstrate.

If a straight line be divided at random, the sum of the rectangles contained by the whole and every one of the parts are equal to the square on the whole.

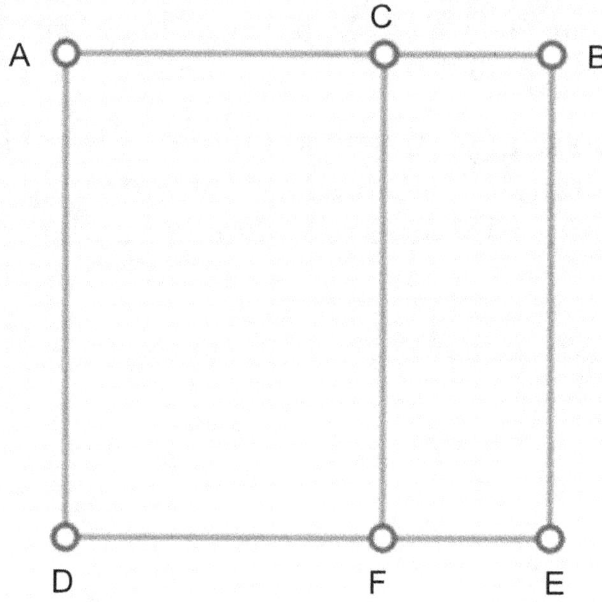

Let there have been given a straight line, AB,
and let it have been divided at random at C.
I say that the sum of the rectangle contained by AB and AC, and the rectangle contained by AB and CB is equal to the square on AB.
Upon the line AB, let there have been described a square, the square ADEB.
And through the point C, let there have been drawn a line parallel to either AD or EB, namely the line CF.
Now, the rectangle AE = the sum of rectangle AF and rectangle CE.
But AE is the square made on AB.
And rectangle AF is the rectangle contained by AB and AC, since it is contained by AC and AD and AD = AB.
And rectangle CE is equal to the rectangle contained by AB and CB, since it is contained by CB and CF, and CF is equal to AD which is equal to AB.
Therefore that which is contained by AB and AC together with that which is contained by AB and BC is equal to the square on AB.

Therefore, if a straight line be divided at random, the sum of the rectangles contained by the whole and every one of the parts is equal to the square on the whole.

The very thing it was required to demonstrate.

Proposition 3

If a straight line be divided at random, the rectangle contained by the whole and one of the parts is equal to: the rectangle contained by the parts and the square of the aforementioned part.

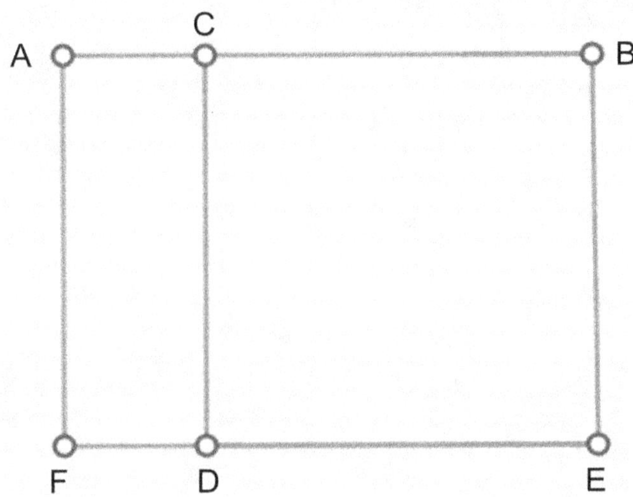

Let there have been given a straight line, AB divided at random at C.
I say that the rectangle contained by AB and BC is equal to: the rectangle contained by AC and BC, and the square on BC.
Upon the line BC, let there have been drawn a square, CDEB.
Extend the line ED.
And through the point A, let there have been drawn a line parallel to either BE or CD, line AF.
(where F is the point where this line meets the extension of ED)
Now the rectangle AE = rectangle AD and rectangle CE.
And AE is the rectangle contained by AB and BC, since it is contained by AB and BE and BE = BC.
And the rectangle AD is the rectangle contained by AC and CB, since DC = CB.
And DB is the square on BC.
Therefore, the rectangle contained by AB and BC is equal to: the rectangle contained by AC and BC, and the square on BC.

Therefore, if a straight line be divided at random, the rectangle contained by the whole and one of the parts is equal to: the rectangle contained by the parts and the square of the aforementioned part.

The very thing it was required to demonstrate.

Proposition 4

If a straight line be cut at random, the square on the whole is equal to the squares
on the parts and twice the rectangle contained by the parts.

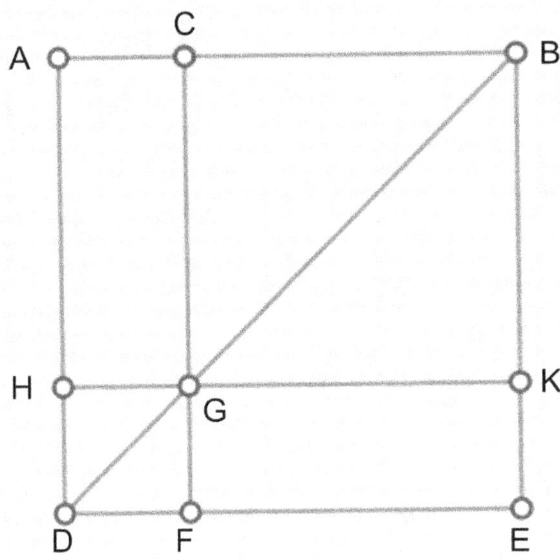

Let there have been a straight line, line AB, and let it have been cut at random at C.
I say that the square on line AB equals: the square on AC, the square on CB, and twice the
rectangle contained by AC and CB.
Let there have been described upon the line AB a square, the square ADEB.
Let the points DB be joined in a straight line.
Through point C, let there have been drawn a straight line parallel to either AD or BE, cutting
the diagonal DB at point G, the line CF.
Through the point G, let there have been drawn a straight line parallel to either AB or DE, the
line HK.
Now since CF is parallel to AD, and upon them falls the line BD, therefore the exterior angle
CGB is equal to the opposite interior angle ADB.
But angle ADB is equal to the angle ABD because BA = AD.
Therefore, angle CGB = angle GBC. Therefore line BC = line CG.
But line CB = line GK and line CG = line KB.
Therefore line GK = line KB.
Therefore the figure CGKB consists of four equal sides.
I say also that it is right angled.
For since CG is parallel to BK, and upon them falls the line straight line CB, the angles KBC and
GCB are equal to two right angles.
But angle KBC is a right angle; therefore, the angle BCG is also a right angle.
Therefore, the angles opposite to them, namely CGK and GKB are right angles.
Therefore CGKB is right angled, and it was proved that the sides are equal; therefore it is a

square and it is described upon the line BC.

And by the same reason, HF is also a square, described upon the line HG, that is, upon the line AC.

Therefore, the squares HF and CK are described upon the lines AC and CB.

And since rectangle AG = rectangle GE, and rectangle AG is contained by AC and CB since CG = CB,

therefore, rectangle GE is equal to that which is contained by AC and CB.

Therefore, rectangles AG and GE are equal to twice that which is contained by AC and CB.

And the squares HF and CK are described upon the lines AC and CB.

Therefore the four rectilinear figures, HF, CK, AG, and GE, are equal to the squares described upon the lines AC and CB and twice the rectangle contained by AC and CB.

But the rectilinear figures, HF, CK, AG, and GE are the whole rectilinear figure ADEB which is the square described upon AB.

Therefore, the square described upon AB is equal to: the square on AC, the square on CB, and twice the rectangle contained by AC and CB.

If therefore, a straight line be cut at random, the square on the whole is equal to the squares on the parts and twice the rectangle contained by the parts.

The very thing it was required to demonstrate.

If a straight line be divided into two equal parts and into two unequal parts, the rectangles contained by the unequal parts of the whole together with the square described on the segment between the cutting points is equal to the square on the half.

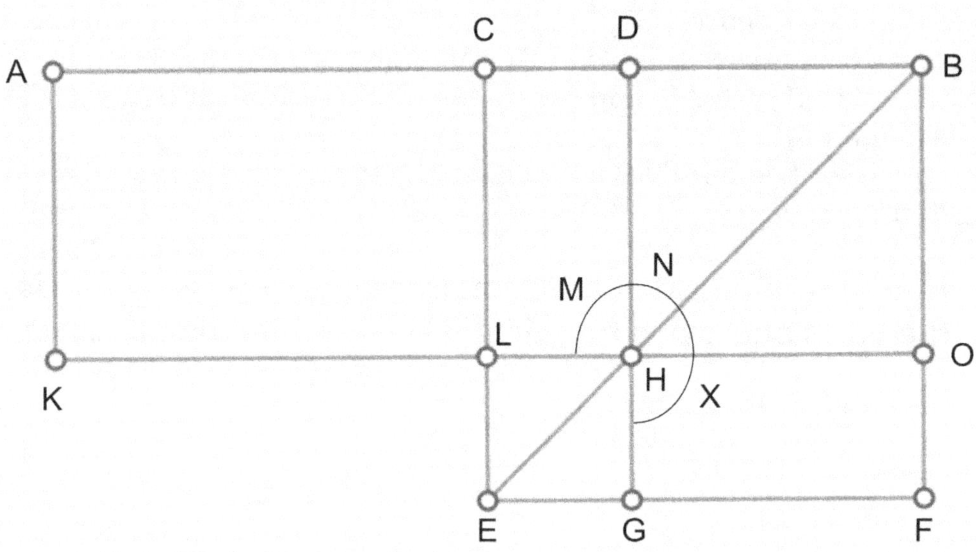

Let there have been a line, AB, and let it have been cut into two equal parts at C, and into unequal parts at D.
I say that the rectangle contained by AD and DB together with the square described on CD is equal to the square on CB.
Upon the line CB, let there have been described the square CEFB.
And let the line EB have been joined.
And through the point D, let there have been drawn a line parallel to CE and BF, cutting the diagonal BE at H, namely line DG.
And again, through the point H, let there have been drawn a line parallel to AB and EF, namely the line KO, (having O being the intersection of KO with BF and letting K be the intersection of KO with the next line to be drawn)
And again, through the point A, let there have been a drawn a line parallel to CL and BO, the line AK.
And since the complement CH = the complement HF, let the parallelogram DO be added to both.
Therefore the whole CO = the whole DF.
But the parallelogram CO = the parallelogram AL, since line AC = line CB.
Therefore, the parallelogram AL = the parallelogram DF.
Let the parallelogram CH be added in common to both.
Therefore, the whole parallelogram AH = parallelogram DL and parallelogram DF.

But parallelogram AH equals that which is contained by AD and DB since DH = DB.
And the parallelograms GH and DF are the gnomon MNX, therefore, the gnomon MNX is equal to that which is contained by AD and DB.
Let the parallelogram LG have been added to both.
Therefore, the gnomon MNX and the parallelogram LG are equal to the rectangle contained by AD and DB, and the square described on CD.
But the gnomon and the parallelogram LG are the whole square CEFB, which is described on BC.
Therefore, the rectangle contained by AD and DB together with the square described on CD is equal to the square on CB.

If therefore a straight line be divided into two equal parts and into two unequal parts, the rectangles contained by the unequal parts of the whole together with the square described on the segment between the cutting points is equal to the square on the half.

The very thing it was required to demonstrate.

If a straight line be divided into two equal parts, and if unto it be added another straight line in a straight line to it, the rectangle contained by the whole of the two lines together and the added line together with the square described on the half of the original line is equal to the square described on the line composed of the half of the original line and the added line.

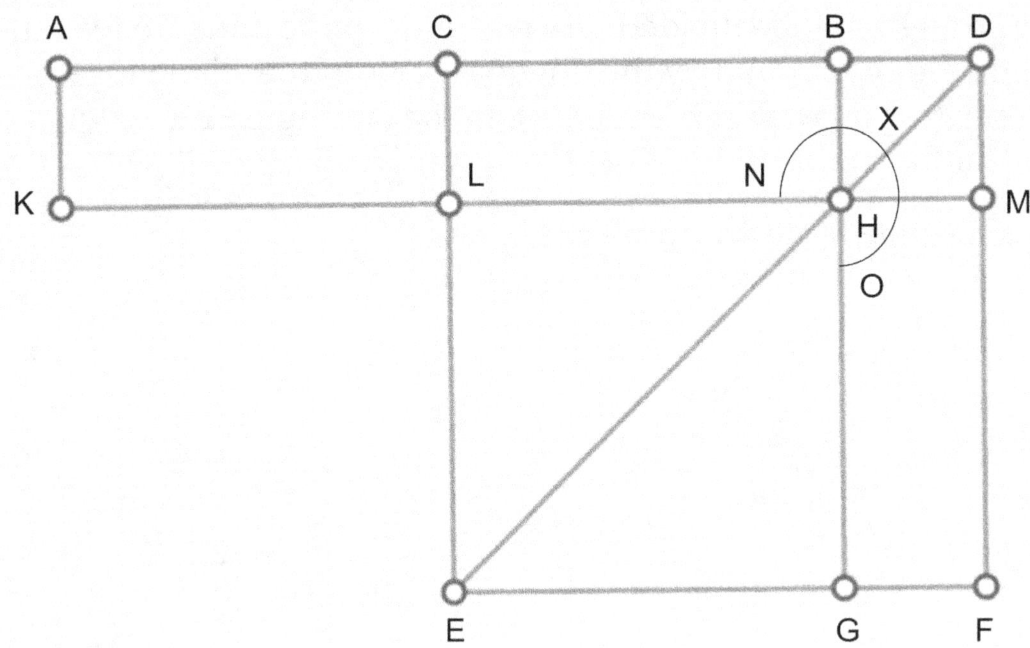

Let there have been given a straight line, line AB, cut into two equal parts at C, and added to it in a straight line the line BD.
I say that the rectangle contained by AD and DB together with the square described on BC equals the square described on DC.
Upon the line DC, let there have been described the square CEFD. And let the line ED have been joined.
And through the point B, let there have been drawn a line parallel to CE and DF, cutting the diagonal ED at H, namely the line BG.
And through point H, let there have been drawn a line parallel to AD and EF, the line KM.
And through the point A, let there have been drawn a line parallel to either line DM or line CL, namely the line AK.
Now since line AC = line CB, parallelogram AL = parallelogram CH.
But parallelogram CH = parallelogram HF, therefore parallelogram AL = parallelogram HF.
Let the parallelogram CM have been added to both.
Therefore the whole AM = the whole gnomon NXO.
But AM is contained by AD and BD since DM = BD.

Therefore the gnomon NXO is equal to the rectangle contained by AD and BD.
Let the parallelogram LG, which is equal to the square on line CB, have been added to both.
Therefore the rectangle contained by AD and BD together with the square on CB is equal to the gnomon NXO and the parallelogram LG.
But the gnomon NXO and the parallelogram LG are the whole square CEFD, which is the square described on DC.
Therefore, the rectangle contained by AD and DB together with the square described on BC equals the square described on DC.

If therefore a straight line be divided into two equal parts, and if unto it be added another straight line in a straight line to it, the rectangle contained by the whole of the two lines together and the added line together with the square described on the half of the original line is equal to the square described on the line composed of the half of the original line and the added line.

The very thing it was required to demonstrate.

If a straight line be divided at random, the square described on the whole together with the square described on one of the parts is equal to twice the rectangle contained by the whole and the one part together with the square of the other part.

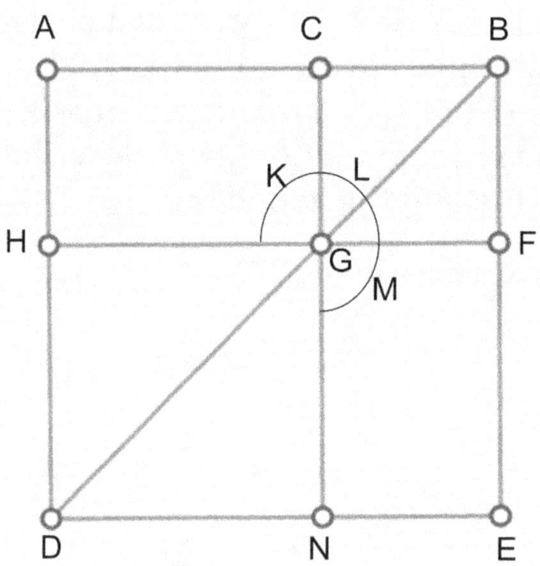

Let there have been a line, AB, divided at random at C.
I say that the square described on AB together with the square described on BC is equal to:
twice the rectangle described by AB and BC and the square on AC.
Upon the line AB, let there have been described a square, square ADEB and complete the figure.
(The figure has the diagonal BD; and line CG parallel to AD, BE where G intersects the diagonal BD; the line HGF drawn through G and parallel to AB, DE; and the gnomon around the square GD labeled KLM).
Now since rectangle AG = rectangle GE, let square CF be added to both.
Therefore, rectangle AF = rectangle CE.
Therefore, rectangles AF and CE together are double of rectangle AF.
But AF and CE together are the gnomon KLM and the square CF.
Therefore, the gnomon KLM and the square CF are double of rectangle AF.
And rectangle AF is that which is contained by AB and BC, since BF=BC.
Therefore, the gnomon KLM and the square CF are double the rectangle contained by AB and BC.
Let the square GD, which is the square on AC, be added to both.
Therefore the gnomon KLM together with the squares CF and GD are equal to: double the rectangle contained by AB and BC together with the square GD.

But the gnomon KLM and the squares BG and GD are the whole square BADE and the square GD, which squares are equal to the squares on AB and BC.
Therefore, the squares on AB and BC are equal to twice the rectangle contained by AB and BC together with the square on AC.

Therefore if a straight line be divided at random, the square described on the whole together with the square described on one of the parts is equal to twice the rectangle contained by the whole and the one part together with the square of the other part.

The very thing it was required to demonstrate.

If a straight line be divided at random, four times the rectangle contained by whole and the one part together with the square on the other part is equal to the square described on the whole and the one part as one line.

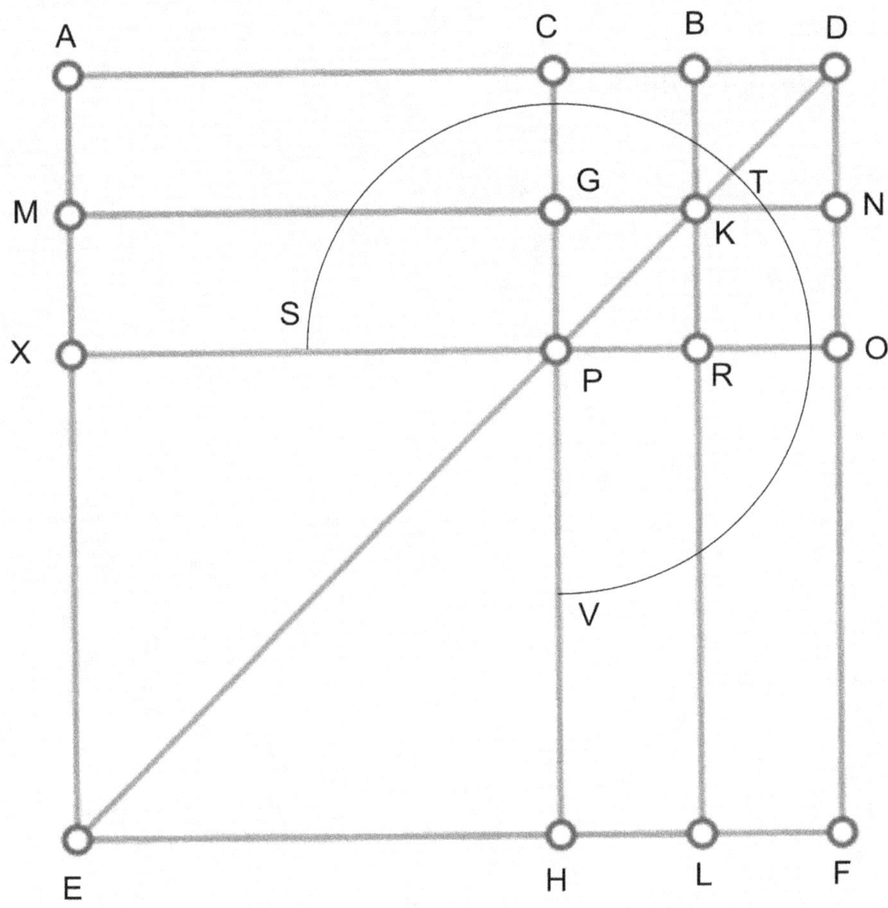

Let there have been a line, AB, divided at random at C.
I say that four times the rectangle contained by AB and BC together with the square on AC is equal to the square described on AB and BC as one line.
Let line AB have been extended past B, and let a line equal to CB have been cut off from the extension, the line BD.
And let a square have been described upon AD, the square AEFD.
And let the line ED have been joined.
And describe a double figure.
Since CB = BD, but CB = GK and likewise BD = KN; therefore, GK = KN.
For the same reason, PR = RO.

Therefore parallelogram CK = parallelogram KD and parallelogram GR = parallelogram RN.
But parallelogram CK = parallelogram RN for they are complements of the parallelogram CO.
Therefore also, parallelogram KD = parallelogram RN.
Therefore the four figures CK, KD, GR, and RN are equal to one another, therefore they are equal to four times CK.
Again, since CB = BD, but BD = BK = CG.
And CB = GK = GP; therefore CG = GP.
And since CG = GP, and PR = RO, the parallelogram AG = parallelogram MP and parallelogram PL = parallelogram RF.
But parallelogram MP = parallelogram PL for they are complements of parallelogram ML.
Therefore also parallelogram AG = parallelogram RF.
Therefore the four figures MP, PL, AG, RF are equal to one another; therefore, the are four times AG.
And it is proved that the four CK, KD, GR, and RN are four times of CK.
Therefore, the eight figures together which contain the gnomon STV are four times the parallelogram AK.
And since parallelogram AK is contained AB and BD since BD = BK; therefore four times that which is contained by AB and BD is four times the parallelogram AK.
And it is proved that the gnomon STV is four times the parallelogram AK.
Therefore, four times that contained by AB and BD = the gnomon STV.
Let the square XH, which is equal to the square described on AC be added to both.
Therefore four times that which is contained by AB and BD together with the square on AC is equal to the gnomon STV together with the parallelogram XH.
But the gnomon STV and the parallelogram XH are the whole square AEFD which was described on line AD.
Therefore, four times that which is contained by AB and BD together with the square on AC is equal to the square on AD.
But BD = BC.
Therefore, four times that which is contained by AB and BD together with the square on AC is equal to the square on AD, that is, on that which is made of AB and BC as one line.

If therefore, a straight line be divided at random, four times the rectangle contained by whole and the one part together with the square on the other part is equal to the square described on the whole and the one part as one line.

The very thing it was required to demonstrate.

If a straight line be divided into two equal parts, and two unequal parts, the squares that are described upon the unequal parts of the whole are double of the squares that are made of the half line and of that line which is between the points of division.

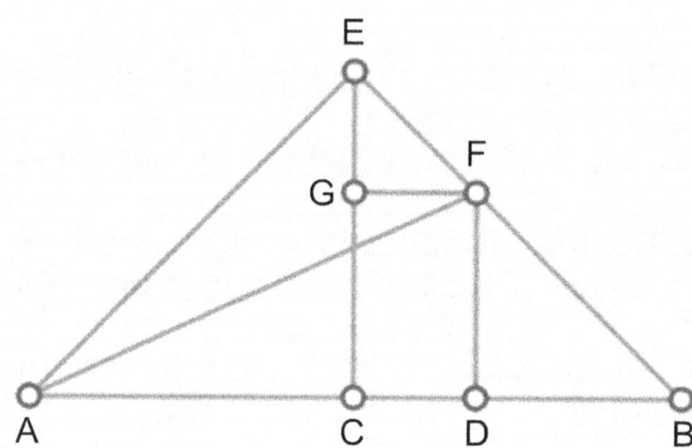

Let there have been a line, AB, and let it have been divided into two equal parts at C, and into two unequal parts at D.
I say that the squares described upon AD and DB are double the squares described upon AC and CD.
From the point C, let there have been raised up a line perpendicular to line AB, the line CE.
And let the point E have been cut off equal to either AC or CB.
And let the lines EA and EB have been joined.
And from point D, let there have been drawn a line parallel to EC, the line DF (where F is the intersection with BE).
And from the point F, let there have been drawn a line parallel to AB, the line FG (where G is the intersection with EC).
And let the line FA have been joined.
Now since AC = CE, angle EAC = angle AEC.
And since the angle a point C is a right angle, the angles remaining, EAC and AEC, are equal to one right angle, therefore, each of the angles CAE and AEC are half of a right angle.
And for the same reason, each of the angles EBC and CEB are half of a right angle.
Therefore, the whole angle AEB is a right angle.
And since GEF is half of a right angle,
but EGF is a right angle because it is equal to the interior opposite angle ECB;
therefore, the remaining angle EFG is half of a right angle.
Therefore, angle GEF = angle EFG; therefore EG = FG.
Again, since the angle at B is half of a right angle, but FDB is a right angle because it is equal to the opposite interior angle ECB; therefore the remaining angle BFD is half of a right angle.
Therefore, the angle at B = angle BFD; therefore, DF = DB.
And since AC = CE, the square described on AC = the square on CE.

Therefore, the squares on AC and CE are double the square on AC.
But the square described on AE = the square on AC and the square on CE because ACE is a right angle.
Therefore the square on AE is double the square on AC.
Again, since EG = GF, the square on EG = the square on GF; therefore the squares on EG and GF are double the square on EG.
But the square on EF equals the squares on EG and GF.
Therefore the square on EF is double the square on EG.
But GF = CD; therefore, the square on EF is double the square on CD.
And the square on AE is double the square on AC.
Therefore the squares on AE and EF are double the squares on AC and CD.
But the square on AF is equal to the squares on AE and EF; therefore, the square on AF is double the squares on AC and CD.
But the squares on AD and DF are equal to the square on AF, for the angle at point D is a right angle.
Therefore, the squares on AD and DF are double the squares on AC and CD.
But DF = DB.
Therefore the squares on AD and DB are double the squares on AC and CD.

If therefore a straight line be divided into two equal parts, and two unequal parts, the squares that are described upon the unequal parts of the whole are double of the squares that are made of the half line and of that line which is between the points of division.

The very thing it was required to demonstrate.

Proposition 10

If a straight line be divided into two equal parts, and to it be added another straight line in a straight line, the square described upon the whole and the added line as one straight line together with the square described upon the added line are double the square described on the half line together with the square described on the half line and the added line as one line.

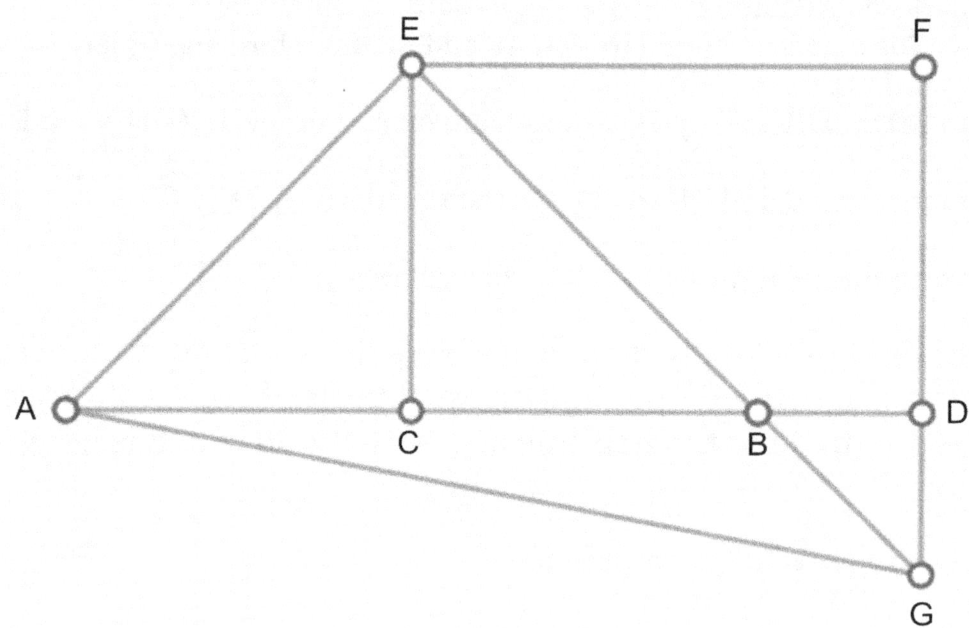

Let there have been a line, AB, and let it have been divided into equal parts at C, and let there have been added to it a straight line in one straight line, namely BD.
I say that the squares described upon AD and DB are double the squares described upon AC and CD.
From the point C, let there have been raised up a line perpendicular to line AB, the line CE.
And let the point E have been cut off equal to either AC or CB.
And let the lines EA and EB have been joined.
And through the point E, let there have been drawn a line parallel to AD, the line EF (where F is the point of intersection between EF and the next line to be drawn).
And through the point D, let there have been drawn a line parallel to CE, the line DF.
And since upon the parallel lines, CE and DF, falls a line EF; therefore, angles CEF and EFD are equal to two right angles.
Therefore, angles FEB and EFD are less than two right angles.
But lines extended indefinitely from angles less than two right angle meet.
Therefore lines EB and FD being extended indefinitely on the side where line BD is will at length meet. Let them have been extended until they meet at a point, point G.
And let AG have been joined.

Now since AC = CE, angle AEC = angle EAC.
And the angle at point C is a right angle; therefore, each of the angles AEC and EAC are half of a right angle.
And for the same reason, each of the angles CEB and EBC are half of a right angle.
Therefore, angle AEB is a right angle.
And since angle EBC is half of a right angle, angle DBG is half of a right angle.
But angle BDG is equal to a right angle because it is equal to the angle DCE for they are alternate angles.
Therefore, the remaining angle DGB is half of a right angle.
Therefore, angle DGB = angle DBG; therefore BD = GD.
And since EGF is half of a right angle, and the angle at point F is a right angle for it is equal to the opposite angle ECD; therefore, the angle remaining, FEG, is equal to half a right angle.
Therefore, angle EGF = angle FEG; therefore, EF = FG.
And since AC = CE, the square described on AC = the square on CE.
Therefore, the squares on AC and CE are double the square on AC.
But the square described on AE = the square on AC and the square on CE.
Therefore the square on AE is double the square on AC.
Again, since GF = EF, the square on GF = the square on EF; therefore the squares on GF and EF are double the squares on EF; but the square on EG = the squares on EF and GF; therefore, the square on EG is double the square on EF.
But EF = CD, therefore, the square on EG is double the square on CD.
And it was proved that the square on AE is double the square on AC.
Therefore, the squares on AE and EG are double the squares on AC and CD.
But the square AG = the squares AE and EG; therefore, the square on AG is double the squares on AC and CD.
But the square on AG = the squares on AD and DG; therefore, the squares on AD and DG are double the squares on AC and CD.
But DG = DB;
therefore the squares on AD and DB are double the squares on AC and CD.

Therefore if a straight line be divided into two equal parts, and to it be added another straight line in a straight line, the square described upon the whole and the added line as one straight line together with the square described upon the added line are double the square described on the half line together with the square described on the half line and the added line as one line.

The very thing it was required to demonstrate.

To divide a straight line in such a way, that the rectangle contained by the whole
and one of the parts is equal to the square described on the other part.

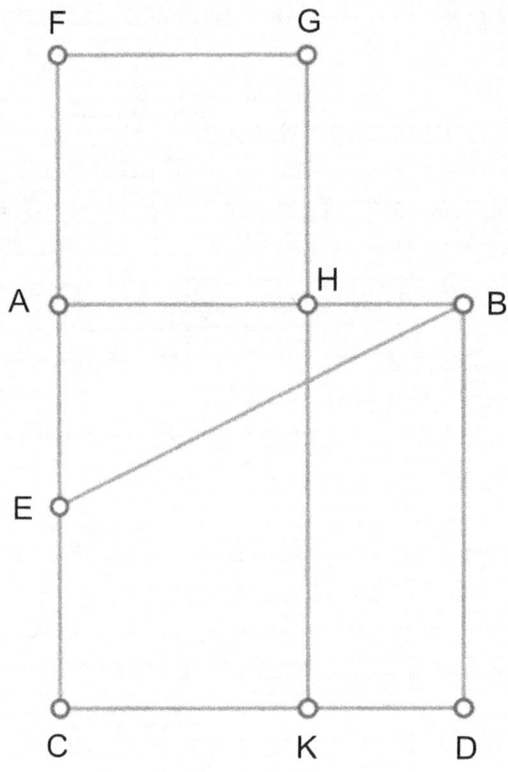

Let there have been given a line, line AB.
It is required to cut line AB in such a way, that the rectangle contained by the whole and one of
the parts is equal to the square described on the other part.
Let a square have been described upon AB, the square ABCD.
And let the line AC have been bisected at point E.
And let BE have been joined.
And extend line EA to F, cutting off EF equal to BE.
And upon the line AF, let a square have been described, the square AFGH.
And extend GH to point K (where K is the intersection with DC.
Then I say that the line AB is divided at point H in such a way that the rectangle contained by AB
and BH is equal to the square on AH.
For since the straight line AC is divided in equal parts at E and another line AF is added to it in a
straight line; therefore, the rectangle contained by CF and FA together with the square on AE is
equal to the square on EF.
But EF = EB; therefore, the rectangle contained by CF and FA together with the square on AE is
equal to the square on EB.
But the square on EB equals the squares on AB and AE; therefore, the rectangle contained by CF
and FA together with the square on AE is equal to the squares AB and AE.

Let the square AE, which is common to both, be taken away; therefore, the rectangle contained by CF and FA is equal to the square AB.

And the rectangle contained by CF and FA is the rectangle FK, for FA = FG.

And the square described on AB is the square AD.

Therefore, figure FK = figure AD.

Let the figure AK, which is common to both, be taken away; therefore, the remainder, figure FH is equal to the remainder, figure HD.

But figure HD is the rectangle contained by AB and BH since AB = BD.

And the figure FH is the square described upon AH.

Therefore, the rectangle contained by AB and BH is equal to the square on AH.

Therefore, the given straight line AB is divided at H in such a way that the rectangle contained by AB and BH is equal to the square described on AH.

The very thing it was required to do.

In obtuse angled triangles, the square described on the side subtending the obtuse angle is greater than the squares described on the sides containing the obtuse angle by twice the rectangle contained by one of the sides containing the obtuse angle, namely that upon which being extended falls a perpendicular line, and the straight line outwardly taken between the obtuse angle and the perpendicular.

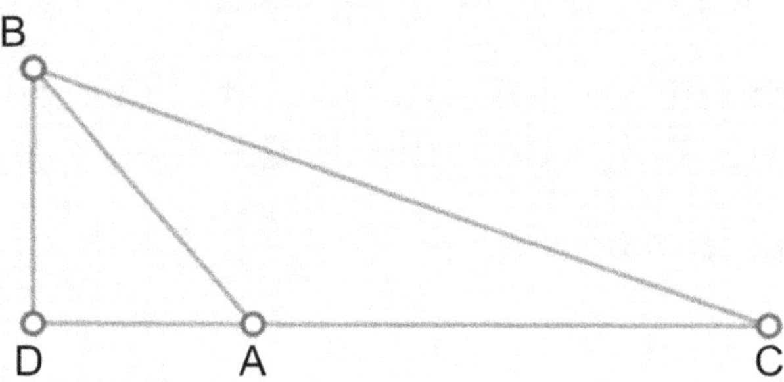

Let there have been given a triangle having an obtuse angle, triangle ABC with the angle BAC being obtuse. And from the point B, let there have been drawn a line perpendicular to CA produced, the line BD.
I say that the square on BC is greater than the squares on BA and AC by twice the rectangle contained by CA and AD.
For since the straight line CD is divided at random at A; therefore the square on CD equals: the squares on CA and AD and twice the rectangle contained by CA and AD.
Let the square DB be added in common to both.
Then the squares on CD and BD equal the squares on CA, AD, and DB together with twice the rectangle contained by CA and AD.
But the square on CB = the squares on CD and BD, since the angle at D is a right angle.
And the squares AD and DB = the square AB.
Therefore, the square on CB is equal to the squares CA and AB together with twice the rectangle contained by CA and AD.

Therefore, in obtuse angled triangles, the square described on the side subtending the obtuse angle is greater than the squares described on the sides containing the obtuse angle by twice the rectangle contained by one of the sides containing the obtuse angle, namely that upon which being extended falls a perpendicular line, and the straight line outwardly taken between the obtuse angle and the perpendicular.

The very thing it was required to demonstrate.

In acute angled triangles, the square on the side subtending the acute angle is less than the squares on the sides containing the acute angle by twice the rectangle contained by one of the sides containing the acute angle, namely that on which the perpendicular falls, and the line inwardly taken between the perpendicular and the acute angle.

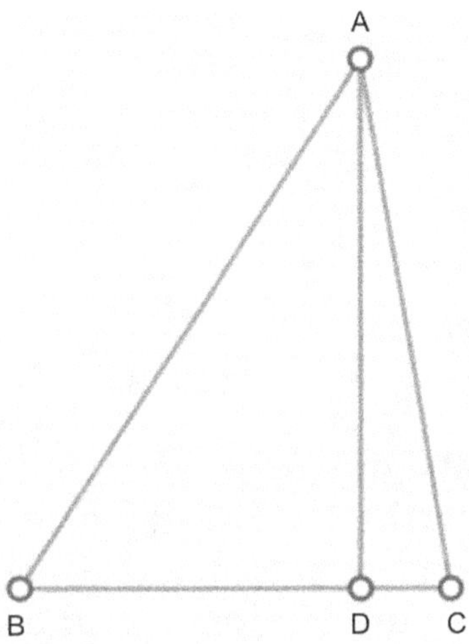

Let there have been given an acute angled triangle, triangle ABC having an acute angle at point B. And through point A, let there have been drawn a line perpendicular to BC, line AD.
I say that the square on AC is less than the squares on AB and BC by twice the rectangle contained by CB and BD.
Since BC is divided at random at D, then the squares on CB and BD are equal to twice the rectangle contained by CB and DB together with the square on CD.
Let the square on DA be added in common to both.
Therefore the squares on CB, DB, and DA = twice the rectangle contained by CB and DB together with the squares on CD and DA.
But the squares on DB and DA = the square on AB, for the angle at point D is a right angle.
And the squares on CD and DA = the square on AC.
Therefore, the squares on CB and AB = the square on AC together with twice the rectangle contained by CB and DB.
Therefore the square on AC alone is less than the squares on CB and AB by twice the rectangle contained by CB and DB.

Therefore in acute angled triangles, the square on the side subtending the acute angle is less than the squares on the sides containing the acute angle by twice the rectangle contained by one of the sides containing the acute angle, namely that on which the perpendicular falls, and the line inwardly taken between the perpendicular and the acute angle.

The very thing it was required to demonstrate.

To describe a square equal to a given rectilineal figure.

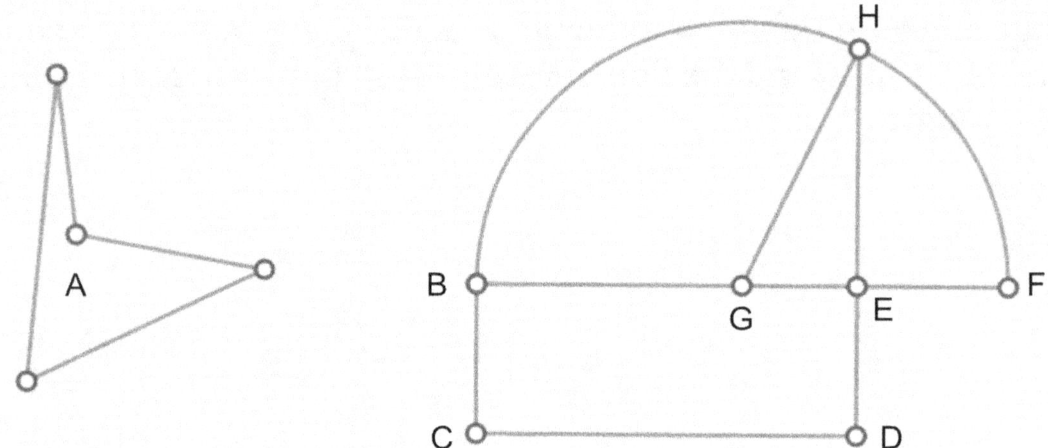

Let there have been given a rectilineal figure A.
It is required to make a square equal to the rectilineal figure A.
Let there have been described a rectangle equal to A, the rectangle BCDE.
Now if BE = ED, then the very thing required is done, for the square BD is equal to the rectilineal figure A.
But if not, one of the lines BE or ED is greater.
Let BE have been the greater, and let it have been extended to F, cutting off F equal to ED.
And let line BF have been bisected at G.
And taking the center G and the distance GB or GF, let a semi-circle have been described, the semi-circle BHF.
And let the line DE have been extended until it hits the semi-circle at H.
And let the line GH have been joined.
Now since the straight line FB is divided into equal parts at G and into unequal parts at E; therefore, the rectangle contained by BE and EF together with the square on EG is equal to the square on GF.
But GF = GH; therefore, the rectangle contained by BE and EF together with the square on EG is equal to the square on GH.
But the square on GH = the squares on HE and EG; therefore, the rectangle contained by BE and EF together with the square on EG is equal to the squares on HE and EG.
Let the square EG which is common to both have been taken away; therefore, the rectangle contained by BE and EF is equal to the square on HE.
But the rectangle contained by BE and EF is the figure BD, since DE = EF.
Therefore the parallelogram BD is equal to the square on HE.
But the parallelogram BD is equal to the rectilineal figure A.
Therefore, the rectilineal figure A is equal to the square on HE.

Therefore a square is described equal to a given rectilineal figure A.

The very thing it was required to do.

Appendix

The Reasons for the Principle Source Publisher Edition of Euclid's Elements

There were two reasons for the writing of the Principle Source Publisher Edition of Euclid's Elements. The first reason is that in the modern age, there is a propensity to value diagrams over words. The second reason is that the terminology chosen by various translators and excess commentary distracts from the simplicity and clarity of Euclid's original work.

The first reason was a pedagogical experiment. The students were asked to draw the diagrams for each proposition without seeing them prior to reading the words. The discipline of drawing a correct diagram from words alone trains students in reasoning and helps avoid the common trap of picture thinking. Since many students will draw many different diagrams, it becomes obvious that the logic of the proposition does not apply to just one diagram, but applies to all possible diagrams. Looking at a completed diagram from the start takes away from the wonder of geometric constructions; for example, even in proposition 1, it is wonderful how given only a line, the bringing out of circles results in the construction of a non-circular equilateral triangle. The given and the goal are more obvious when a student draws a diagram, because he or she must draw the given first before drawing the goal. The experiment of requiring the student to draw all diagrams herself or himself was only partly successful. For beginning learners, a new practice of having every student direct the drawing of proposition 2 blindfolded replaced the use of a book with no diagrams.

For study with more than one participant, the best practice is to have less advanced or more advanced learners present and redraw the proposition diagram without looking at the diagram in the book. In the Geometry and Reasoning classes at Magdalen College of the Liberal Arts (Bedford and then Warner, New Hampshire), tutors have always recognized that to avoid memorization without understanding, when presenting a proposition, the presenter does not need to label the points in the diagram as they are in the book. At times, instead of letters, the points are given names (point Bob, point Jack, point Sue…) or colors, or some other alternate name. If some one of the listeners says, "But Euclid said line AB = CD, not DF = CA," that interlocutor will be directed to follow the presentation with the letters as written on the board now and not disturb the line of logic with arbitrary remembrances.

The second reason for an adaptation of the original translation is more difficult to explain, and perhaps controversial. Principle Source Publisher Edition of Euclid's Elements is closer to the original Greek. This claim is made not because the person writing the adaptation knows Greek better than Sir Thomas Heath, but because of the willful choice to state literally what Euclid wrote rather than what the translator thinks Euclid meant. The way in which this is true can only be explained by outlining two views of translation.

The first view of translation is that an honored expert, who knows intimately the original language, the subject, and the new language writes a translation of what a work means. Knowing the mindset and history of the ancient writer, he translates the meaning of what was written into modern terms that modern people may understand what was written even though they do not know the intricacies, mindset, and history of the ancient writer. To do this, the honored expert can not translate word for word. The ancient language and the modern language are too different: adjustments must be made for the modern reader to understand the content of the ancient writer's work. The honored expert also has read centuries worth of commentaries and is aware of the progress of modern ideas in the field. As a label, this first view of translation is called here the meaning for meaning translation.

The second view of translation is that regardless of the difficulties presented to the modern reader, language is universal, and the best translation of an ancient work is the one that is closest to a word for word translation. In this view, the translator does his best to preserve as much of the order and content of the original as possible. The goal is to present in the modern language what the author said and leave it up to the modern reader to decide what the writer means. As a label, this second view of translation is called word for word translation.

The epitome of meaning for meaning translation applied to Euclid's Elements is obviously the edition prepared by Sir Thomas Heath. An honored mathematician, expert on the history of mathematics, and Greek scholar, he prepared a lucid, easily read, and fully complete and accurate translation of Euclid's Elements with copious notes and aids to the reader.

The epitome of word for word translation applied to Euclid's Elements does not exist and as far as I know, will not exist in the foreseeable future. It would take a scholar equal to Sir Thomas Heath who held the view of word for word translation to produce such a translation.

Inbetween is the Principle Source Publisher Edition of Euclid's Elements. It follows the word for word view of translation. In following that view, it is closer to the original Greek than other available translations.

This can be shown clearly by examining the notes found in the edition of Sir Thomas Heath, and comparing Heath's translation to the Principle Source Publisher Edition. Heath, as a good and honest scholar, himself indicates where he chose not to follow the "literal" translation, but came up with a slightly different translation that he claims to be more understandable in English:

> *From any point to any point.* In general statements of this kind, the Greeks did not say, as we do, "*any* point," "*any* triangle" etc., but "*every* point," "*every* triangle," and the like. Thus the words are here literally "from every point to every point." Similarly, the first words of Postulate 3 are "with *every* centre and distance," and the enunciation, e.g., of I.18 is "In every triangle the greater side subtends the greater angle." Heath pg. 195

> *Let AB be the given straight line.* To be strictly literal we should have to translate in the reverse order, "let the given finite straight line be the (straight line) AB"; but this order is inconvenient in other cases where there is more than one datum, e.g., in the setting-out of I.2, "let the given point be A, and the given straight line BC," the awkwardness arising from the omission of the verb in the second clause. Hence I have, for clearness sake, adopted the other order throughout the book. (Heath pg. 242)

> *Let the circle BCD be described.* Two things are here to be noted, (1) the elegant and practically universal use of the perfect passive imperative in constructions, *gegpaxtho*, meaning of course "let it *have been* described" or "suppose it described" , (2) the impossibility of expressing shortly in a translation the force of the words in their original order…. means literally, "let a circle have been described, the (circle, namely,

which I denote by) BCD." Similarly we have lower down, "let straight lines, (namely) the (straight lines) CA, CB, be joined," There seems to be no practicable alternative, in English, but to translate as I have done in the text. (Heath pg. 242)

Heath's notes clearly divides how "we" say things from how "the Greeks" say things: "the Greeks did not say, as we do…" This seems to assume, what I believe to be an explanatory fiction, that there is a difference between the "Greek mind," and "our mind." It also implies that the primitive Greek mind was further from the truth of things than the enlightened modern mind. If translations must be adapted to the current mind of peoples, is Heath's translation any good for Americans? Heath is an English aristocrat and is not American. Shouldn't Americans have a new American version of Euclid rather than read Heath's English aristocratic version?

But since Euclid has been translated and appreciated down the centuries in Latin, Spanish, French, and the English of the times of Henry Billingsley, perhaps the theory that there is no difference between "the Greek mind" and any person's mind from any time deserves consideration.

It remains to describe this editor's adaptations used to produce the Principle Source Publisher Edition of Euclid's Elements.

The first English translation of the Elements by Billingsley was chosen because Billingsley, though a Greek scholar, did not think of himself as primarily a mathematician. His translation follows the Greek word order very closely, and not being a mathematician he did not try to improve or correct Euclid. Unfortunately, some of the English used by Billingsley sounds as odd as Greek to modern readers, so the first set of adaptations was to use modern geometrical terms for Billingsley's older geometric terms. For example, what Billingsley calls "right lines" is adapted to modern usage as "straight lines."

The second set of adaptations comes directly from Heath's notes. In cases where Heath (or Billingsley) chose not to translate literally, the Principle Source Publisher Edition does translate literally. Billingsley also chose to use "any" instead of "every", and he translated the past perfect imperative usually as "suppose it to be" whereas the "let it have been" seemed more literally the sense of the Greek.

The third set of adaptations comes from the corrected Greek text compiled by Heiberg that was the basis of Heath's edition. Between the time when Billingsley made the first English translation and Heath made his translation, older Greek texts of Euclid were discovered. In fact, Heath refers to Heiberg's new Greek edition as a primary reason why a new English edition was needed.

The fourth set of adaptations was to remove from the text all notes, editor's helps (like references to previous propositions), and editorial additions. It seems Euclid expected the student to remember and to recognize the use of previous propositions without hints in the text.

The first edition was printed without diagrams, and constructions that are unclear without a diagram have helpful descriptions within parentheses that seemed necessary to help students draw the diagram. It is obvious that Euclid intended to have diagrams in The Elements since at one point the words say "complete the diagram," without giving specific instructions. In this second edition, diagrams are provided to help beginners and those reading alone.

In conclusion, the Principle Source Publisher Edition of Euclid is closer to the original Greek through choice. The editor does not claim greater knowledge of Greek than other translators. The editor does not claim greater insight into Mathematics or the History of Mathematics than any other translator of Euclid. But in a way, this is an advantage, since the editor has no other way to improve the edition other than to try to say what Euclid said in the way that he said it, and to let Euclid's genius recommend this edition.